Specifications for identity and purity of food colours

As prepared by the 28th session of the
Joint FAO/WHO expert committee on food additives

Rome, 19-28 March 1984

FOOD AND NUTRITION PAPER

31/1

JECFA

FOOD
AND
AGRICULTURE
ORGANIZATION
OF THE
UNITED NATIONS
Rome, 1984

The designations employed and the presentation of material in this publication do not imply the expression of any opinion whatsoever on the part of the Food and Agriculture Organization of the United Nations concerning the legal status of any country, territory, city or area or of its authorities, or concerning the delimitation of its frontiers or boundaries.

M-84
ISBN 92-5-102127-9

All rights reserved. No part of this publication may be reproduced, stored in a retrieval system, or transmitted in any form or by any means, electronic, mechanical, photocopying or otherwise, without the prior permission of the copyright owner. Applications for such permission, with a statement of the purpose and extent of the reproduction, should be addressed to the Director, Publications Division, Food and Agriculture Organization of the United Nations, Via delle Terme di Caracalla, 00100 Rome, Italy.

© FAO 1984

TABLE OF CONTENTS

	Page
INTRODUCTION	v
LIST OF PARTICIPANTS	vii

MONOGRAPHS

Allura Red AC	3
Aluminium Powder	7
Amaranth	9
β-Apo-8'-Carotenal	13
β-Apo-8'-Carotenoic Acid Ethyl Ester	17
Azorubine	21
Beet Red	23
Brilliant Black BN	27
Brilliant Blue FCF	31
Brown HT	35
Canthaxanthin	39
Carbon Blacks	43
Carotenes - Natural	45
β-Carotene - Synthetic	47
Carthamus Red	51
Carthamus Yellow	55
Chlorophylls	57
Chlorophyll-Copper Complex	59
Citrus Red No 2	61
Cochineal and Carminic Acid	63
Curcumin	67
Erythrosine	71
Fast Green FCF	77
Fast Red E	81
Ferrous Gluconate	83
Grape Skin Extract	87
Green S	91
Indigotine	95
Iron Oxides	99
Lithol Rubine BK	101
Paprika Oleoresin	105

Page

MONOGRAPHS (cont'd)

Patent Blue V 111
Ponceau 4R 115
Quinoline Yellow 117
Red 2G 121
Riboflavin 123
Riboflavin 5'-Phosphate Sodium 127
Sunset Yellow FCF 131
Tartrazine 133
Titanium Dioxide 135
Xanthophylls 141

General Specifications for Aluminium Lakes
of Colouring Matters 143

LIST OF FOOD COLOURS FOR WHICH THE EXISTING
SPECIFICATIONS HAVE BEEN WITHDRAWN 145

ANNEX - METHODS OF ANALYSIS

Indexed Separately

INTRODUCTION

The specifications for identity and purity appearing in this publication were prepared at the Twenty-eighth Session of the Joint FAO/WHO Expert Committee on Food Additives (Rome, 19-28 March 1984) for substances which, in the consideration of the Committee, adequate data were available. Readers are required to consider these specifications only in conjunction with the Report of the above-mentioned meeting, WHO Technical Report Series (in press).

Part 1 of this publication covers food colours, which were reviewed separately by the Committee. Part 2 covers the other substances considered.

The General principles governing the elaboration of specifications have been summarized in Annex 4 of the Tenth Report.* The specifications for identity and purity of food additives established by JECFA are meant to identify the substance that has been subject to biological testing, to ensure that the substance is of adequate degree of purity required for safe use in food and to reflect and encourage good manufacturing practices. They are established mainly for the use of toxicologists and others concerned with identity and purity of the substance. As agreed by the Committee at its Twenty-sixth Session, specifications may also be established prior to the eventual completion of toxicological evaluation, in certain cases when the available toxicological data are inadequate or incomplete and do not permit the establishment of full or temporary ADIs. References are made in individual specifications to some of the criteria that may be of interest in commerce, but they do not necessarily include all requirements of interest to the commercial user. These specifications are not more stringent than is necessary to accomplish their purpose and should easily be attainable by the producing industries. The report of the Twenty-third Session gives the reasons why certain specifications are designated "tentative".

Attention is drawn to the following sections of the Twenty-eighth Report: principles governing the toxicological evaluation of compounds on the agenda (Section 2.2); principles governing the establishment and revision of specifications (Section 2.3), namely "the need for sufficient data" (Section 2.3.1), "review of specifications for food colours" (Section 2.3.2); comments on specific food additives and contaminants (Section 3); and establishment, revision and withdrawal of certain specifications (Section 4).

The general methods and the test solutions referred to in the specifications are those appearing in Guide to Specifications (General notices, General methods, Identification tests, Test solutions, other reference materials), FAO Food and Nutrition Paper No 5, Revision 1 (1983).

Summaries of information on the toxicological evaluation of food additives considered at the Twenty-eighth Session will be available from the World Health Organization, Geneva.

Information requested on individual specifications should be communicated to the Food Quality and Consumer Protection Group, Food Quality and Standards Service, Food Policy and Nutrition Division, FAO, 00100 Rome, Italy.

---

* Tenth Report, FAO Nutrition Meetings Report Series No 43 (1967); WHO Tech. Rep. Series No 373 (1967).

JOINT FAO/WHO EXPERT COMMITTEE ON FOOD ADDITIVES
Twenty-eighth Session
Rome, 19-28 March 1984

List of Participants

Members invited by WHO

Dr H. Blumenthal, Director, Division of Toxicology, Bureau of Foods, Food and Drug Administration, Washington DC, USA

Professor E. Fournier, Professor of Clinical Toxicology, Hôpital Fernand-Widal, Paris, France

Dr G. Nazario, Rua Monte Alegre, Sao Paulo, Brazil

Dr P. Pothisiri, Director, Technical Division, Food and Drug Administration, Ministry of Public Health, Bangkok, Thailand

Professor M.J. Rand, Head, Department of Pharmacology, University of Melbourne, Victoria, Australia (Vice-Chairman)

Dr A.N. Zaicev, Head, Laboratory for Hygienic Research of Food Additives, Institute of Nutrition, Academy of Medical Sciences of the USSR, Moscow, USSR

Members invited by FAO

Dr W.H.B. Denner, Principal Scientific Officer, Food Science Division, Ministry of Agriculture, Fisheries and Food, London, England

Dr S.W. Gunner, Director, Bureau of Chemical Safety, Food Directorate, Health Protection Branch, Department of National Health and Welfare, Ottawa, Canada (Chairman)

Professor K. Kojima, College of Environmental Health, Azabu University, Sagamihara-Shi, Kanagawa, Japan

Dr J.P. Modderman, Food Additive and Animal Drug Chemistry, Department of Health and Human Sciences, Food and Drug Administration, Washington DC, USA

Professor F.J. Pellerin, Faculty of Pharmacy, University of Paris XI, Centre hospitalier Corentin-Celton, Issy-les-Moulineaux, France

Observers invited by FAO

Mr A. Feberwee, Chairman, Codex Committee on Food Additivies, Ministry of Agriculture and Fisheries, The Hague, Netherlands

Dr R. Mathews, Director, Food Chemicals Codex, National Academy of Sciences, Washington DC, USA

Secretariat

Dr C.L. Galli, Professor of Experimental Toxicology, University of Milan, Milan, Italy (WHO Temporary Adviser)

Mr R. Haigh, Principal Administrator, Commission of the European Communities, Brussels, Belgium (WHO Temporary Adviser)

Dr Y. Hayashi, Director, Division of Pathology, Biological Safety Research Centre, National Institute of Hygienic Sciences, Tokyo, Japan (WHO Temporary Adviser)

Dr J.E. Long, Head, Toxicological Evaluation Division, Health Protection Branch, Department of National Health and Welfare, Ottawa, Canada (WHO Temporary Adviser)

Dr B. MacGibbon, Senior Principal Medical Officer, Division of Toxicology, Environmental Pollution and Prevention, Department of Health and Social Security, London, England (Rapporteur)

Professor I. Nir, Department of Pharmacology and Experimental Therapeutics, Hebrew University, Hadassah Medical School, Jerusalem, Israel (WHO Temporary Adviser)

Mr A. Pearce, Director of Technical Services, Williams (Hounslow) Limited, Middlesex, England (FAO Consultant)

Dr A.W. Randell, Food Policy and Nutrition Division, FAO, Rome, Italy (FAO Joint Secretary)

Dr N. Rao Maturo, Food Standards Officer, Joint FAO/WHO Food Standards Programme, Food Policy and Nutrition Division, FAO, Rome, Italy

Dr S.I. Shibko, Associate Director for Regulatory Evaluation, Division of Toxicology, Bureau of Foods, Food and Drug Administration, Washington DC, USA (WHO Temporary Adviser)

Dr P. Shubik, Resident Fellow, Green College, Oxford, England (WHO Temporary Adviser)

Professor K. Topsy, Dr Molière Street, Beau Bassin, Mauritius (WHO Temporary Adviser)

Dr G. Vettorazzi, Toxicologist, International Programme on Chemical Safety, Division of Environmental Health, WHO, Geneva, Switzerland (WHO Joint secretary)

Dr R. Walker, Department of Biochemistry, University of Surrey, Guildford, England (WHO Temporary Adviser)

Mr J. Wilbourn, Unit of Carcinogen Identification and Evaluation, International Agency for Research on Cancer, Lyon, France (WHO Temporary Adviser)

MONOGRAPHS

ALLURA RED AC*

SYNONYMS	CI Food Red 17 FD&C Red No.40
DEFINITION	Allura Red AC consists essentially of disodium 2-hydroxy-1-(2-methoxy-5-methyl-4-sulfonatophenylazo) naphthalene-6-sulfonate and subsidiary colouring matters together with sodium chloride and/or sodium sulfate as the principal uncoloured components. Allura Red AC may be converted to the corresponding aluminium lake in which case only the General Specifications for Aluminium Lakes of Colouring Matters shall apply.
Class	Monoazo
Code numbers	CI (1975) No.16035 CI (1975) Food Red 17 CAS No.25956-17-6
Chemical name	Disodium 2-hydroxy-1-(2-methoxy-5-methyl-4-sulfonatophenylazo) naphthalene-6-sulfonate
Chemical formula	$C_{18}H_{14}N_2Na_2O_8S_2$
Structural formula	

$$NaO_3S\text{-}\underset{CH_3}{\underset{|}{\overset{OCH_3}{\overset{|}{C_6H_2}}}}\text{-}N=N\text{-}\underset{SO_3Na}{\overset{HO}{C_{10}H_5}}$$

Molecular weight	496.42
Assay	Content not less than 85% total colouring matters
DESCRIPTION	Dark red powder or granules
FUNCTIONAL USE	Food colour
CHARACTERISTICS	

IDENTIFICATION TESTS

**A. Solubility	Soluble in water Insoluble in ethanol

---

* Supersedes the earlier specifications for Allura Red AC published in FAO Food and Nutrition Paper No 19 (1981).

** See General Methods (Guide to Specifications, FAO Food and Nutrition Paper No 5, Revision 1, 1983).

IDENTIFICATION TESTS (cont'd)

*B. Identification of colouring matters

PURITY TESTS

*Loss on drying at 135°	Not more than 15%
*Chloride and sulfate calculated as sodium salts	Not more than 15%
*Water insoluble matter	Not more than 0.2%
Subsidiary colouring matters	Not more than 3% (See description under TESTS)

Organic compounds other than colouring matters

    6-hydroxy-2-naphthalene sulfonic acid, sodium salt:
        Not more than 0.3%
        (See description under TESTS)

    4-amino-5-methoxy-2-methylbenzene sulfonic acid:
        Not more than 0.2%
        (See description under TESTS)

    6,6-oxybis (2-naphthalene sulfonic acid) disodium salt:
        Not more than 1.0%
        (See description under TESTS)

    Unsulfonated primary aromatic amines
        Not more than 0.01% calculated as aniline

*Ether extractable matter	Not more than 0.2%
*Arsenic	Not more than 3 mg/kg
*Lead	Not more than 10 mg/kg
**Heavy metals	Not more than 40 mg/kg

TESTS

    PURITY TESTS

    *Subsidiary colouring matters

    Use the following conditions:
      Developing solvent: No. 4
      Height of ascent of solvent front: approximately 17 cm

    *Organic compounds other than colouring matters

    Use HPLC under the following conditions:
    HPLC elution gradient:
        0 to 18% at 1% per minute (linear)
        then 18 to 62% at 7% per minute (linear)
        followed by elution at 100%

    Flow rate: 0.6 ml per minute

---

*   See Annex.

** See General Methods (Guide to Specifications, FAO Food and Nutrition Paper No 5, Revision 1, 1983).

METHOD OF ASSAY

*Determination of Total Colouring Matters by Titration with Titanous Chloride

Use the following:

Amount to weigh: 0.5 - 0.6 g
Buffer: 15 g sodium hydrogen tartrate
Weight (D) of colouring matters equivalent to 1.00 ml of
0.1 N $TiCl_3$: 0.01241 g

---

* See Annex.

ALUMINIUM POWDER *
(Tentative)**

DEFINITION	Aluminium powder is composed of finely divided particles of aluminium. The grinding may or may not be carried out in the presence of edible vegetable oils and/or food additive quality fatty acids.** It is free from admixture with substances other than edible vegetable oils and/or food additive quality fatty acids.
Code numbers	CI (1975) No 77000 CI (1975) Pigment Metal 1 EEC No E 173
Chemical name	Aluminium
Chemical formula	Al
Atomic weight	26.98
Assay	Not less than 99% calculated as Al on an oil-free basis.
DESCRIPTION	A silvery grey powder
FUNCTIONAL USE	Food colour; decorative pigment

CHARACTERISTICS

IDENTIFICATION TESTS

***A. Solubility	Insoluble in water Insoluble in organic solvents Soluble in dilute hydrochloric acid
B. Identification	A solution in dilute hydrochloric acid gives the reactions characteristic of Al^{3+}.

PURITY TESTS

****Loss on drying at 105°C	Not more than 0.5%
****Arsenic	Not more than 3 mg/kg
****Lead	Not more than 20 mg/kg
***Heavy metals	Not more than 40 mg/kg

---

* Supersedes the earlier specification for Aluminium published in FAO Nutrition Meeting Report Series No 57 (1977).

** Information required on methods of grinding.

*** See General Methods (Guide to Specifications, FAO Food and Nutrition Paper No 5, Revision 1, 1983).

****See Annex.

METHOD OF ASSAY   Transfer about 0.2 g of the sample, accurately weighed, to a 500-ml flask fitted with a rubber stopper carrying a 150-ml separating funnel, an inlet tube connected to a cylinder of carbon dioxide and an outlet tube dipping into a water-trap. Add 60 ml of freshly boiled and cooled water and disperse the sample; replace the air by carbon dioxide and add, by the separating funnel, 100 ml of a solution containing 56 g of ferric ammonium sulfate and 7.5 ml of sulfuric acid in freshly boiled and cooled water. While maintaining an atmosphere of carbon dioxide in the flask, heat to boiling and boil for 5 minutes. After the sample has dissolved, cool rapidly to 20°, and dilute to 250 ml with freshly boiled and cooled water. To 50 ml of this solution, add 15 ml of phosphoric acid and titrate with 0.1 N potassium permanganate.

1 ml of 0.1 N potassium permanganate = 0.8994 mg Al.

AMARANTH *

SYNONYMS	CI Food Red 9; Naphtol Rot S.
DEFINITION	Amaranth consists essentially of trisodium 2-hydroxy-1-(4-sulfonato-1-naphthylazo) naphthalene-3,6-disulfonate and subsidiary colouring matters together with sodium chloride and/or sodium sulfate as the principal uncoloured components. Amaranth may be converted to the corresponding aluminium lake in which case only the General Specifications for Aluminium Lakes of Colouring Matters shall apply.
Class	Monoazo
Code numbers	CI (1975) No 16185 CI (1975) Food Red 9 CAS No 915-67-3 EEC No E 123
Chemical name	Trisodium 2-hydroxy-1-(4-sulfonato-1-naphthylazo) naphthalene-3,6-disulfonate
Chemical formula	$C_{20}H_{11}N_2Na_3O_{10}S_3$
Structural formula	

Molecular weight	604.48
Assay	Content not less than 85% total colouring matters
DESCRIPTION	Reddish brown to dark reddish brown powder or granules
FUNCTIONAL USE	Food colour

---

* Supersedes the earlier specifications for Amaranth published in FAO Food and Nutrition Paper No 25 (1982).

CHARACTERISTICS

   IDENTIFICATION TESTS

   *A. Solubility   Soluble in water
                              Sparingly soluble in ethanol

   **B. Identification of colouring matters

   PURITY TESTS

   **Loss on drying at 135°   Not more than 15%

   **Chloride and sulfate
      calculated as sodium salts   Not more than 15%

   **Water insoluble matter   Not more than 0.2%

   Subsidiary colouring matters   Not more than 3%
                                      (See description under TESTS)

   Organic compounds other than colouring matters

         4-aminonaphthalene-1-sulfonic acid
         3-hydroxynaphthalene-2,7-disulfonic acid   Total not more
         6-hydroxynaphthalene-2-sulfonic acid   than 0.5%
         7-hydroxynaphthalene-1,3-disulfonic acid   (See description
         7-hydroxynaphthalene-1,3,6-trisulfonic acid   under TESTS)

   ** Unsulfonated primary aromatic amines:
                                      Not more than 0.01%
                                      calculated as aniline

   **Ether extractable matter   Not more than 0.2%

   **Arsenic   Not more than 3 mg/kg

   **Lead   Not more than 10 mg/kg

   *Heavy metals   Not more than 40 mg/kg

TESTS

   PURITY TESTS

   **Subsidiary colouring matters

      Use the following conditions:

         Developing solvent: No 3
         Height of ascent of solvent front: 17 cm, then 1 hour further
                                            development

---

*   See General Methods (Guide to Specifications, FAO Food and Nutrition
    Paper No 5, Revision 1, 1983).
** See Annex.

TESTS (cont'd)

*Organic compounds other than colouring matters

Use HPLC under the following conditions:

HPLC elution gradient: 2 to 100% at 4.0% per minute (linear)

METHOD OF ASSAY

*Determination of total colouring matters by titration with titanous chloride

Use the following:

Weight of sample: 0.7 - 0.8 g
Buffer: 10 g sodium citrate
Calculation: Weight (D) of colouring matters equivalent to 1.00 ml of 0.1 N $TiCl_3$: 0.01511 g

---

* See Annex

β-APO-8'-CAROTENAL*

SYNONYM	CI Food Orange 6
DEFINITION	
Class	Carotenoid
Code numbers	CI (1975) No 40820 CI (1975) Food Orange 6 CAS No 1107-26-2 EEC No E 160e
Chemical names	β-apo-8'-carotenal; 8'-Apo-β-caroten-8'-al.
Chemical formula	$C_{30}H_{40}O$
Structural formula	
Molecular weight	416.65
Assay	β-apo-8'-carotenal is predominantly the trans isomer and contains not less than 96% of total colouring matters.
DESCRIPTION	Deep violet crystals with metallic lustre or crystalline powder.
	β-apo-8'-carotenal is sensitive to oxygen and light and should therefore be kept in a light-resistant container under inert gas. Users of β-apo-8'-carotenal generally prefer to purchase the stabilized forms that are commercially available rather than the product that assays at not less than 96%. These stabilized forms are prepared from β-apo-8'-carotenal, meeting the specification set out below, and include solutions or suspensions of β-apo-8'-carotenal in edible fats or oils, emulsions and water dispersable powders. These preparations may contain permitted antioxidants, emulsifying agents and dispersing agents and may have different trans/cis isomer ratios from the parent colour described in the specification. The analytical methods prescribed for the parent colour are not necessarily suitable for the assay of, or determination of impurities in the stabilized forms.
FUNCTIONAL USE	Food colour

---

* Supersedes the earlier specification for β-apo-8'-carotenal published in FAO Food and Nutrition Paper No 19 (1981).

CHARACTERISTICS

IDENTIFICATION TESTS

*A. Solubility   Insoluble in water
Slightly soluble in ethanol
Sparingly soluble in vegetable oils
Soluble in chloroform

B.   Determine the absorbance of the sample solution (see Method of Assay) at 461 nm and 488 nm. The ratio A_{488}/A_{461} is between 0.80 and 0.84.

C.   Passes test
(See description under TESTS)

D.   Passes test
(See description under TESTS)

PURITY TESTS

<u>Sulfated ash</u>   Not more than 0.1%
(See description under TESTS)

<u>Subsidiary colouring matters</u>

Carotenoids other than β-apo-8'-carotenal:
Not more than 3%
(See description under TESTS)

**<u>Arsenic</u>   Not more than 3 mg/kg

**<u>Lead</u>   Not more than 10 mg/kg

*<u>Heavy metals</u>   Not more than 40 mg/kg

TESTS

IDENTIFICATION TESTS

C.   The colour of a solution of the sample in acetone disappears after the successive additions of a 5% solution of sodium nitrite and 1 N sulfuric acid.

D.   A solution of the sample in chloroform turns blue on addition of an excess of Carr-Price reagent TS.

PURITY TESTS

*<u>Sulfated ash</u>   Amount to weigh: 2.0 g

---

*   See General Methods (Guide to Specifications, FAO Food and Nutrition Paper No 5, Revision 1, 1983).

**   See Annex.

PURITY TESTS (cont'd)

Subsidiary colouring matters

Carotenoids other than β-apo-8'-carotenal:

Dissolve about 80 mg of β-apo-8'-carotenal in 100 ml chloroform. Apply 400 μl of this solution as streak 2 cm from the bottom of the thin-layer plate (0.25 mm silicagel).

Pretreat the thin-layer plate by soaking in a tank with 3% KOH in methanol so that it is completely wetted. Then dry the plate for 5 minutes in the air and activate for 1 hour at 110°C in an oven. Let cool over $CaCl_2$ and keep in a desiccator over $CaCl_2$.

Immediately after applying the carotenoid solution to the plate, develop the chromatogram with n-hexane/chloroform/ethylacetate (70+20+10) in a saturated chamber suitably protected from light, until the solvent front has moved 10 cm above the initial streak. Remove the plate, allow the main part of the solvent to evaporate at room temperature and mark the principal band as well as the bands corresponding to other carotenoids. Remove the silica gel from the principal band, transfer it to a glass-stoppered 100-ml centrifuge tube and add 40.0 ml of chloroform.
(Solution 1)

Separately remove the silica gel of the combined bands corresponding to other carotenoids, transfer to a glass-stoppered 50-ml centrifuge tube and add 20.0 ml chloroform.
(Solution 2)

Shake the centrifuge tubes by mechanical means for 10 minutes and centrifuge for 5 minutes. Dilute 10.0 ml of solution 1 to 50.0 ml with chloroform.
(Solution 3)

Determine, with a suitable spectrophotometer, the absorbances of Solutions 2 and 3 in 1 cm cells at the wavelength of maximum in chloroform at about 474 nm, using chloroform as blank.

Calculation:

Carotenoids other than β-apo-8'-carotenal (%) =

$$\frac{A_2 \cdot 10}{A_3}$$

where A_2 = absorbance of Solution 2; and
A_3 = absorbance of Solution 3.

./.

*METHOD OF ASSAY

Weigh accurately about 0.08 g of the sample and proceed as directed in the spectrophotometric method in the Annex.

absorbtivity (a) = 2640

Approximate wavelength of maximum absorption = 461 nm.

---

* See Annex

β-APO-8'-CAROTENOIC ACID ETHYL ESTER *

SYNONYMS CI Food Orange 7
β-Apo-8'carotenoic acid ethyl ester
Ethyl 8'-apo-β-ψ-caroten-8'-oate

DEFINITION

    Class Carotenoid

    Code numbers CI (1975) No 40825
CI (1975) Food Orange 7
CAS No 1109-11-1
EEC No E 160f

    Chemical names β-Apo-8'-carotenoic acid ethyl ester;
Ethyl-8'-apo-β-ψ-caroten-8'-oate.

    Chemical formula $C_{32}H_{44}O_2$

    Structural formula

    Molecular weight 460.70

    Assay β-apo-8'-carotenoic acid ethyl ester is predominantly trans isomer and contains not less than 96% total colouring matters.

DESCRIPTION Red to violet-red crystals or crystalline powder.

β-apo-8'-carotenoic acid ethyl ester is sensitive to oxygen and light and should therefore be kept in a light-resistant container under inert gas. Users of β-apo-8'-carotenoic acid ethyl ester generally prefer to purchase the stabilized forms that are commercially available rather than the product that assays at not less than 96%. These stabilized forms are prepared from β-apo-8'-carotenoic acid ethyl ester meeting the specification set out below and include solutions or suspensions of β-apo-8'-carotenoic acid ethyl ester in edible fats or oils, emulsions and water dispersable powders. These preparations may contain permitted antioxidants, emulsifying agents and dispersing agents and may have different trans/cis isomer ratios from the parent colour described in the specification. The analytical methods prescribed for the parent colour are not necessarily suitable for the assay of, or determination of impurities in the stabilized forms.

FUNCTIONAL USE Food colour

---

* Supersedes the earlier specifications for β-apo-8'-carotenoic acid ethyl ester published in FAO Food and Nutrition Paper No 19 (1981).

CHARACTERISTICS

IDENTIFICATION TESTS

*A. Solubility   Insoluble in water
                      Very slightly soluble in ethanol
                      Slightly soluble in vegetable oils
                      Soluble in chloroform

B.   Determine the absorbance of the sample solution
                      (see Method of Assay) at 449 nm and 475 nm. The
                      ratio A_{475}/A_{449} is between 0.82 and 0.86.

C.   Passes test
                      (See description under TESTS)

D.   Passes test
                      (See description under TESTS)

PURITY TESTS

Sulfated ash   Not more than 0.1%
                      (See description under TESTS)

Subsidiary colouring matters
    Carotenoids other than β-apo-8'carotenoic acid ethyl ester:
    Not more than 3%
    (See description under TESTS)

**Arsenic   Not more than 3 mg/kg

**Lead   Not more than 10 mg/kg

*Heavy metals   Not more than 40 mg/kg

TESTS

IDENTIFICATION TESTS

C.   The colour of a solution of the sample in acetone
                      disappears after successive additions of a 5%
                      solution of sodium nitrite and 1 N sulfuric acid.

D.   A solution of the sample in chloroform turns blue
                      on addition of an excess of Carr-Price reagent TS.

PURITY TESTS

Sulfated ash   Weigh accurately 2.0 g and proceed as directed
                      under "Sulfated Ash" in the General Methods.*

---

*   See General Methods (Guide to Specifications, FAO Food and Nutrition
     Paper No 5, Revision 1, 1983).

** See Annex.

PURITY TESTS (cont'd)

Subsidiary colouring matters

Carotenoids other than β-apo-8'-carotenoic acid ethyl ester:

Dissolve about 80 mg of β-apo-8'-carotenoic acid ethyl ester in 100 ml chloroform. Apply 400 μl of this solution as a streak 2 cm from the bottom of the thin-layer plate (Silicagel 0.25 mm).

Pretreat the thin-layer plate by soaking in a tank with 3% KOH in methanol so that it is completely wetted. Dry the plate then for 5 minutes in the air and activate for 1 hour at 110°C in an oven. Let cool over $CaCl_2$ and keep in a desiccator over $CaCl_2$.

Immediately after applying the carotenoid solution to the plate, develop the chromatogram with n-hexane/toluene/di-isopropyl-ether (70+20+10) in a saturated chamber suitably protected from light, until the solvent front has moved 15 cm above the initial streak.

Remove the plate, allow the main part of the solvent to evaporate at room temperature and mark the principal band as well as the bands corresponding to other carotenoids. Remove the silica gel from the principal band, transfer it to a glass-stoppered 100-ml cetrifuge tube and add 40.0 ml of chloroform.
(Solution 1)

Separately remove the silica gel of the combined bands corresponding to other carotenoids, transfer to a glass-stoppered 50-ml centrifuge tube and add 20.0 ml chloroform.
(Solution 2)

Shake the centrifuge tubes by mechanical means for 10 minutes and centrifuge for 5 minutes. Dilute 10.0 ml of Solution 1 to 50.0 ml with chloroform.
(Solution 3)

Determine, with a suitable spectrophotometer, the absorbances of Solutions 2 and 3 in 1 cm cells at the wavelength of maximum in chloroform at about 455 nm, using chloroform as blank.

Calculation:

Carotenoids other than β-apo-8'-carotenoic acid ethyl ester (%) =
$$\frac{A_2 \cdot 10}{A_3}$$
where A_2 = absorbance of Solution 2; and
A_3 = absorbance of Solution 3.

*METHOD OF ASSAY   Weigh accurately about 0.08 g of the sample and proceed as directed in the spectrophotometric method in the Annex.

absorbtivity (a) = 2550

Approximate wavelength of maximum absorption = 449 nm.

---

* See Annex.

AZORUBINE*

SYNONYMS	CI Food Red 3 Carmoisine
DEFINITION	Azorubine consists essentially of disodium 4-hydroxy-3-(4-sulfonato-1-naphthylazo) naphthalene-1-sulfonate and subsidiary colouring matters together with sodium chloride and/or sodium sulfate as the principal uncoloured components. Azorubine may be converted to the corresponding aluminium lake in which case only the General Specifications for Aluminium Lakes of Colouring Matters shall apply.
Class	Monoazo
Code numbers	CI (1975) No 14720 CI (1975) Food Red 3 CAS No 3567-69-9 EEC No E 122
Chemical name	Disodium 4-hydroxy-3-(4-sulfonato-1-naphthylazo) naphthalene-1-sulfonate.
Chemical forumla	$C_{20}H_{12}N_2Na_2O_7S_2$
Structural formula	
Molecular weight	502.44
Assay	Content not less than 85% total colouring matters
DESCRIPTION	Red powder or granules
FUNCTIONAL USE	Food colour

CHARACTERISTICS

IDENTIFICATION TESTS

  **A. Solubility   Soluble in water
                               Sparingly soluble in ethanol

 ***B. Identification of colouring matters

---

*   Supersedes the earlier specification for Azorubine published in FAO Food and Nutrition Paper No 19 (1981).

** See General Methods (Guide to Specifications, FAO Food and Nutrition Paper No 5, Revision 1, 1983).

*** See Annex.

CHARACTERISTICS (cont'd)

PURITY TESTS

*Loss on drying at 135°	Not more than 15%
*Chloride and sulfate calculated as sodium salts	Not more than 15%
*Water insoluble matter	Not more than 0.2%
Subsidiary colouring matters	Not more than 1% (See description under TESTS)

Organic compounds other than colouring matters

    4-aminonaphthalene-1-sulfonic acid ⎫ Total not more than 0.5%
    4-hydroxynaphthalene-1-sulfonic acid ⎭ (See description under TESTS)

    Unsulfonated primary aromatic amines:
        Not more than 0.01% calculated as aniline

*Ether extractable matter	Not more than 0.2%
*Arsenic	Not more than 3 mg/kg
*Lead	Not more than 10 mg/kg
**Heavy metals	Not more than 40 mg/kg

TESTS

PURITY TESTS

*Subsidiary colouring matters

    Use the following conditions:

        Developing solvent: No 4
        Height of ascent of solvent front: approximately 17 cm

*Organic compounds other than colouring matters

    Use HPLC under the following conditions:

        HPLC elution gradient: 1 to 100% at 2.0% per minute (exponential)

METHOD OF ASSAY

*Determination of Total Colouring Matters by Titration with Titanous Chloride

    Use the following:

        Weight of sample: 0.5 - 0.6 g
        Buffer: 15 g sodium hydrogen tartrate
        Weight (D) of colouring matters equivalent to 1.00 ml of 0.1 N $TiCl_3$: 0.01256 g

---

* See Annex.
** See General Methods (Guide to Specifications, FAO Food and Nutrition Paper No 5, Revision 1, 1983).

BEET RED*
(Tentative)

SYNONYM	Beetroot Red
DEFINITION	Beet red is obtained from the roots of red beets (*Beta vulgaris* L. var *rubra*) as press juice or by aqueous extraction of shredded beetroots. The solution may be further concentrated or formed into paste, powder or solid of which the principal pigment is Betanine. During manufacture of some products most of the salts and/or sugars and proteins naturally present are removed and food grade acids (e.g. citric, lactic, L-ascorbic) may be added as pH controlling agents and stabilizers.
Class	Betalain
Code number	EEC No. E162
**Chemical name	Beet red contains Betanine as the principal pigment which is the D-glucopyronoside of Betanidine.
Chemical formula	Betanine: $C_{24}H_{26}N_2O_{13}$
**Structural formula	Betanine is:

---

* Supersedes the earlier specifications for Beet Red published in FAO Food and Nutrition Paper No 25 (1982).

At the time of revising these specifications, the previously allocated temporary ADI "not specified" was withdrawn.

** Additional information required regarding the stereochemistry of glucose linkage and on levels of arsenic and betanine in commercial products.

Molecular weight | Betanine: 550.48

*Assay | Beet red contains not less than 1.0% in liquid form; not less than 4% in powder form (calculated as Betanine).

DESCRIPTION | Red to dark red liquid, concentrate, paste, powder or solid

FUNCTIONAL USE | Food colour

CHARACTERISTICS

IDENTIFICATION TESTS

**A. Solubility | Soluble in or miscible with water
Insoluble in or immiscible with absolute ethanol

B. Colour reaction | Addition of a sodium hydroxide solution (1 in 10) to an aqueous solution of Beet Red changes the colour from red to reddish violet to yellow.

C. Spectrophotometry | Betanine in water at pH 5.4 has an absorbance maximum at about 530 nm and at pH 8.9 exhibits a broadened maximum at about 545 nm.

D. Thin layer chromatography
(a) On cellulose plates (0.25 mm) with Sørensen's phosphate buffer (pH 5.6) as solvent, Beet Red colour gives a number of spots in various colours (yellow, orange, red, purple, violet). Betanine appears as a purple spot with an Rf value of about 0.7.

Sørensen's phosphate buffer (pH 5.6):

Solution A: 1/15 $\underline{M}$ potassium hydrogen phosphate: Dissolve 9.08 g of KH_2PO_4 in water and dilute to 1000 ml.

Solution B: 1/15 $\underline{M}$ disodium hydrogen phosphate: Dissolve 11.88 g of $Na_2HPO_4 \cdot 2H_2O$ in water and dilute to 1000 ml.

pH 5.6: 94.8 ml of solution A + 5.2 ml of solution B.

(b) On cellulose plates (0.10 mm) in the solvent (2 g sodium citrate + 78.5 ml water + 21.5 ml concentrated ammonia water) betanine follows the front of the liquid as distinct from acidic water-soluble synthetic dyes. In this liquid betanine is yellow.

---

* Additional information required on Betanine content of commercial products.

** See General Methods (Guide to Specifications, FAO Food and Nutrition Paper No 5, Revision 1, 1983).

PURITY TESTS

> Basic dyes
>
> Passes test.
>
> To 1 g of the sample add 100 ml of 1% sodium hydroxide solution, and mix well. Extract 30 ml of this solution with 15 ml of ether. When extracted wash the ether layer twice with dilute acetic acid (5 ml); the dilute acetic acid layer does not produce a colour.
>
> *Other acidic dyes
>
> Passes test.
>
> To 1 g of the sample add 1 ml of ammonia TS and 8 ml of water, and shake well. Discard an oily layer when separated. Proceed as directed under Paper Chromatography (Ascending Chromatography) in General Methods*** using 2 μl of the solution as the sample solution, and a mixture of pyridine and ammonia TS (2:1 by volume) as the developing solvent. Stop the development when the solvent front has advanced about 15 cm from the point of application. No spot is observed at the solvent front after drying under daylight. If any spot is observed, it should be decolourized when sprayed with a solution of stannous chloride (2 in 5).
>
> **Arsenic
>
> Not more than 3 mg/kg
>
> ***Lead
>
> Not more than 10 mg/kg
>
> ***Heavy metals
>
> Not more than 40 mg/kg
>
> Test 0.5 g of the sample as directed in the Limit Test (Method II) using 20 μg of lead ion (Pb) in the control (Solution A).

METHOD OF ASSAY

> Reagents and Solutions
>
> McIlvaine's citric acid phosphate buffer (pH 5)
>
> Solution A: 0.1 $\underline{M}$ citric acid. Dissolve 21.0 g of citric acid ($C_6H_8O_7 \cdot 1H_2O$) in water and dilute to 1000 ml.
>
> Solution B: 0.2 $\underline{M}$ disodium hydrogen phosphate. Dissolve 35.6 g of $Na_2HPO_4 \cdot 2H_2O$ in water and dilute to 1000 ml.
>
> pH 5.0: 48.5 ml of solution A + 51.5 ml of solution B.
>
> Procedure
>
> Dissolve a quantity of Beet Red accurately weighed (W g, 4 decimals), in McIlvaine's citric acid phosphate buffer (pH 5) and dilute to a suitable volume with the buffer solution (V ml in total); the

./.

---

* Additional information required regarding the adequacy of this test.

** Additional information required on the level of Arsenic in commercial products.

*** See General Methods (Guide to Specifications, FAO Food and Nutrition Paper No 5, Revision 1, 1983).

METHOD OF ASSAY (cont'd)

Procedure (cont'd)

maximum absorption shall be within the range of 0.2 to 0.8. Centrifuge the solution if necessary, and measure the absorption with water as reference. The colour intensity is calculated on the basis of the maximum absorption A (at about 530 nm), all red colouring matter being included under betanine with specific extinction

$E_{1cm}^{1\%} = 1120.$

Calculation

$$\% \text{ Beet Red} = \frac{A \times V}{1120 \times L \times W}$$

in which A = maximum absorption
V = volume of test solution measured in ml
L = length of cell measured in cm
W = weight of sample in g.

BRILLIANT BLACK BN *

SYNONYMS	CI Food Black 1 Black PN
DEFINITION	Brilliant Black BN consists essentially of tetrasodium 4-acetamido-5-hydroxy-6-[7-sulfonato-4-(4-sulfonato phyenylazo)-1-naphthylazo] naphthalene-1,7-disulfonate and subsidiary colouring matters together with sodium chloride and/or sodium sulfate as the principal uncoloured components. Brilliant Black BN may be converted to the corresponding aluminium lake in which case only the General Specifications for Aluminium Lakes of Colouring Matters apply.
Class	Bisazo
Code numbers	CI (1975) No 28840 CI (1975) Food Black 1 CAS No 2519-30-4 EEC No E 151
Chemical name	Tetrasodium 4-acetamido-5-hydroxy-6-[7-sulfonato-4-(4-sulfonatophenylazo)-1-naphthylazo] naphthalene-1,7-disulfonate
Chemical formula	$C_{28}H_{17}N_5Na_4O_{14}S_4$
Structural formula	$NaO_3S-C_6H_4-N=N-\text{(naphthyl)}-N=N-\text{(naphthyl, OH, NHCOCH}_3\text{, SO}_3Na\text{, SO}_3Na\text{)}$
Molecular weight	867.69
Assay	Content not less than 80% total colouring matter
DESCRIPTION	Black powder or granules
FUNCTIONAL USE	Food colour
CHARACTERISTICS	

IDENTIFICATION TESTS

**A. Solubility	Soluble in water Sparingly soluble in ethanol

---

* Supersedes the earlier specifications for Brilliant Black BN published in FAO Food and Nutrition Paper No 19 (1981).

** See General Methods (Guide to Specifications, FAO Food and Nutrition Paper No 5, Revision 1, 1983).

IDENTIFICATION TESTS (cont'd)

*B. Identification of colouring matters

PURITY TESTS

*Loss on drying at 135°	Not more than 20%
*Chloride and sulfate calculated as sodium salts	Not more than 20%
*Water insoluble matter	Not more than 0.2%
Subsidiary colouring matters	Not more than 4% (See description under TESTS)

Organic compounds other than colouring matters

- Sum of 4-acetamido-5-hydroxynaphthalene-1,7-disulfonic acid
- and 4-amino-5-hydroxynaphthalene-1,7-disulfonic acid
- 8-aminonaphthalene-2-sulfonic acid
- Sulfanilic acid
- 4,4' diazoaminodi-(benzene sulfonic acid)

Not more than 0.8% (See description under TESTS)

*Unsulfonated primary aromatic amines:	Not more than 0.01% calculated as aniline
*Ether extractable matter	Not more than 0.2%
*Arsenic	Not more than 3 mg/kg
*Lead	Not more than 10 mg/kg
**Heavy metals	Not more than 40 mg/kg

TESTS

PURITY TESTS

*Subsidiary colouring matters

Use the following conditions:

Developing solvent: Chromatogram (i): No 1.
Chromatogram (ii): No 4.
Height of ascent of solvent front:
(i): approximately 17 cm
(ii): approximately 17 cm

---

*   See Annex.

** See General Methods (Guide to Specifications, FAO Food and Nutrition Paper No 5, Revision 1, 1983).

PURITY TESTS (cont'd)

*Organic compounds other than colouring matters

    Use HPLC under the following conditions:

        HPLC elution gradient: 2 to 100% at 2% per minute (linear)

METHOD OF ASSAY

*Determination of Total Colouring Matters by Titration with Titanous Chloride

    Use the following:

        Weight of sample: 0.6-0.7 g
        Buffer: 15 g sodium hydrogen tartrate
        Weight (D) of colouring matters equivalent to
            1.00 ml of 0.1 N $TiCl_3$: 0.01086 g

---

* See Annex

BRILLIANT BLUE FCF *

SYNONYMS	CI Food Blue 2 FD&C Blue No 1
DEFINITION	Brilliant Blue FCF consists essentially of disodium α-[4-(N-ethyl-3-sulfonatobenzylamino) phenyl]-α-[4-(N-ethyl-3-sulfonatobenzyliminio) cyclohexa-2,5-dienylidene] toluene-2-sulfonate and its isomers and subsidiary colouring matters together with sodium chloride and/or sodium sulfate as the principal uncoloured components. Brilliant Blue FCF may be converted to the corresponding aluminium lake in which case only the General Specifications for Aluminium Lakes of Colouring Matters shall apply.
Class	Triarylmethane
Code numbers	CI (1975) No 42900 CI (1975) Food Blue 2 CAS No 3844-45-9 EEC No E 133
Chemical name	Disodium α-[4-(N-ethyl-3-sulfonatobenzylamino) phenyl]-α-[4-(N-ethyl-3-sulfonatobenzyliminio) cyclohexa-2,5-dienylidene] toluene-2-sulfonate.
Chemical formula	$C_{37}H_{34}N_2Na_2O_9S_3$
Structural formula	
Molecular weight	792.84
Assay	Content not less than 85% total colouring matters
DESCRIPTION	Blue powder or granules
FUNCTIONAL USE	Food Colour
CHARACTERISTICS	

IDENTIFICATION TESTS

**A. Solubility	Soluble in water Slightly soluble in ethanol

---

* Supersedes the earlier specifications for Brilliant Blue FCF published in FAO Nutrition Meeting's Report Series No 38B (1966).

** See General Methods (Guide to Specifications, FAO Food and Nutrition Paper No 5, Revision 1, 1983).

IDENTIFICATION TESTS (cont'd)

*B. Identification of colouring matters

PURITY TESTS

*Loss on drying at 135°	Not more than 15%
*Chloride and sulfate calculated as sodium salts	Not more than 15%
*Water insoluble matter	Not more than 0.2%
Subsidiary colouring matters	Not more than 6% (See description under TESTS)

Organic compounds other than colouring matters

    Sum of 2-, 3- and 4-formyl benzene sulfonic acids:
        Not more than 1.5%
        (See description under TESTS)

    3-[[ethyl](4-sulfophenyl)amino]methyl benzene sulfonic acid:
        Not more than 0.3%
        (See description under TESTS)

Leuco base:	Not more than 5%
*Unsulfonated primary aromatic amines:	Not more than 0.01% calculated as aniline
*Ether extractable matter	Not more than 0.2%
*Arsenic	Not more than 3 mg/kg
*Lead	Not more than 10 mg/kg
*Chromium	Not more than 50 mg/kg
**Heavy metals	Not more than 40 mg/kg

TESTS

PURITY TESTS

*Subsidiary colouring matters

    Use the following conditions:

        Developing solvent: No 4
        Develop chromatogram for approximately 20 hours

---

*   See Annex

** See General Methods (Guide to Specifications, FAO Food and Nutrition Paper No 5, Revision 1, 1983).

PURITY TESTS (cont'd)

*Organic compounds other than colouring matters

Proceed as directed under Column Chromatography

The following absorptivities may be used:

3-formyl benzene sulfonic acid: 0.0495 mg/L/cm at 246 nm in dilute HCl.

3-[[ethyl](4-sulfophenyl)amino] methyl benzene sulfonic acid: 0.078 mg/L/cm at 277 nm in dilute ammonia.

*Leuco base

Weigh accurately 120 ± 5 mg of sample and proceed as directed under "Determination of Leuco Base."

Absorptivity (a) = 0.164 mg/L/cm at approximately 630 nm
Ratio = 0.9706

METHOD OF ASSAY

*Determination of Total Colouring Matters by Titration with Titanous Chloride

Use the following:

Weight of sample: 1.8 - 1.9 g
Buffer: 15 g sodium hydrogen tartrate
Weight (D) of colouring matters equivalent to
1.00 ml of 0.1 N $TiCl_3$: 0.03965 g

---

*   See Annex

BROWN HT *

SYNONYMS	CI Food Brown 3 Chocolate brown HT
DEFINITION	Brown HT consists essentially of disodium 4,4'-(2,4-dihydroxy-5-hydroxymethyl-1,3-phenylene bisazo) di(naphthalene-1-sulfonate) and subsidiary colouring matters together with sodium chloride and/or sodium sulfate as the principal uncoloured components. Brown HT may be converted to the corresponding aluminium lake in which case only the General Specifications for Aluminium Lakes of Colouring Matters shall apply.
Class	Bisazo
Code numbers	CI (1975) No 20285 CI (1975) Food Brown 3 CAS No 4553-89-3 EEC No 156
Chemical name	Disodium 4,4'-(2,4-dihydroxy-5-hydroxymethyl-1,3-phenylene bisazo)di(naphthalene-1-sulfonate)
Chemical formula	$C_{27}H_{18}N_4Na_2O_9S_2$
Structural formula	NaO_3S—[naphthalene]—N=N—[phenyl with OH, HO, CH_2OH]—N=N—[naphthalene]—SO_3Na
Molecular weight	652.57
Assay	Content not less than 70% total colouring matters
DESCRIPTION	Brown powder or granules
FUNCTIONAL USE	Food colour

CHARACTERISTICS

IDENTIFICATION TESTS

**A. Solubility   Soluble in water
                    Insoluble in ethanol

***B. Identification of colouring matters

---

* Supersedes the earlier specification for Brown HT published in FAO Food and Nutrition Paper No 25 (1982).

** See General Methods (Guide to Specifications, FAO Food and Nutrition Paper No 5, Revision 1, 1983).

*** See Annex.

PURITY TESTS

*Loss on drying at 135°	Not more than 30%
*Chloride and sulfate calculated as sodium salts	Not more than 30%
*Water insoluble matter	Not more than 0.2%
Subsidiary colouring matters	Not more than 10% (See description under TESTS)

Organic compounds other than colouring matters

    4-aminonaphthalene-1-sulfonic acid:
        Not more than 0.7%
        (See description under TESTS)

    *Unsulfonated primary aromatic amines:
        Not more than 0.01%
        calculated as aniline

*Ether extractable matter	Not more than 0.2%
*Arsenic	Not more than 3 mg/kg
*Lead	Not more than 10 mg/kg
**Heavy metals	Not more than 40 mg/kg

TESTS

    PURITY TESTS

    *Subsidiary colouring matters

        Use the following conditions:

        Prepare the standard in the following manner:
        Dilute 1.50 ml of the 1% dye solution to 100 ml with water and mix well. Transfer 0.10 ml of this solution to a test tube; add 5.0 ml of water : acetone (1:1 by vol.) and then 14.9 ml of 0.05 N sodium hydrogen carbonate solution and shake the tube to ensure mixing. Determine the net absorbance (A_s) of the standard.

            Developing solvent: No 6
            Develop chromatogram for approximately 14 hours.

    *Organic compounds other than colouring matters

        Use HPLC under the following conditions:

        HPLC elution gradient: 1 to 100% at 2.0% per minute (exponential)

---

*   See Annex.

**   See General Methods (Guide to Specifications, FAO Food and Nutrition Paper No 5, Revision 1, 1983).

METHOD OF ASSAY

Determination of Total Colouring Matters by Spectrophotometry

Use the following conditions:

Solvent: pH7 phosphate buffer
Dilution of solution A: 10 ml ⟶ 250 ml
Absorptivity (a): 40.3
Approximate wavelength of maximum absorption: 460 nm

---

* See Annex.

CANTHAXANTHIN *

SYNONYM	CI Food Orange 8
DEFINITION	
Class	Carotenoid
Code numbers	CI (1975) No 40850 CI (1975) Food Orange 8 CAS No 514-78-3 EEC No E 161g
Chemical names	β-Carotene-4,4'-dione Canthaxanthin 4,4'-Dioxo-β-carotene
Chemical formula	$C_{40}H_{52}O_2$
Structural formula	
Molecular weight	564.86
Assay	Canthaxanthin is predominantly trans isomer and contains not less tha 96% of total colouring matters.
DESCRIPTION	Deep violet crystals or crystalline powder.

Canthaxanthin is sensitive to oxygen and light and should therefore be kept in a light-resistant container under inert gas. Users of Canthaxanthin generally prefer to purchase the stabilized forms that are commercially available rather than the product that assays at not less then 96%. These stabilized forms are prepared from Canthaxanthin meeting the specification set out below and include solutions or suspensions of Canthaxanthin in edible fats or oils, emulsions and water dispersable powders. These preparations may contain permitted antioxidants, emulsifying agents and dispersing agents and may have different trans/cis isomer ratios from the parent colour described in the specification. The analytical methods prescribed for the parent colour are not necessarily suitable for the assay of, or determination of impurities in the stabilized forms.

FUNCTIONAL USE   Food Colour

---

* Supersedes the earlier specifications for Canthaxanthin published in FAO Food and Nutrition Paper No 19 (1981).

CHARACTERISTICS

IDENTIFICATION TESTS

*A. Solubility   Insoluble in water
                                  Insoluble in ethanol
                                  Practically insoluble in vegetable oils
                                  Soluble in chloroform

B.   A solution of canthaxanthin in cyclo-
                                  hexane has an absorbance maximum between
                                  468 and 472 nm.

C.   Passes test
                                  (See description under TESTS)

D.   Passes test
                                  (See description under TESTS)

PURITY TESTS

Sulfated ash   Not more than 0.1%
                                  (See description under TESTS)

Subsidiary colouring matters

    Carotenoids other than Canthaxanthin:
                                  Not more than 5%
                                  (See description under TESTS)

**Arsenic   Not more than 3 mg/kg

**Lead   Not more than 10 mg/kg

*Heavy metals   Not more than 40 mg/kg

TESTS

IDENTIFICATION TESTS

C.   The colour of a solution of canthaxanthin in
                                  acetone disappears after successive additions
                                  of a 5% solution of sodium nitrite and 1 N sul-
                                  furic acid.

D.   A solution of the sample in chloroform turns
                                  blue on addition of an excess of Carr-Price
                                  reagent TS.

PURITY TESTS

Sulfated ash   Amount to weigh: 2.0 g

---

*   See General Methods (Guide to Specifications, FAO Food and Nutrition Paper No 5, Revision 1, 1983).

** See Annex.

PURITY TESTS (cont'd)

Subsidiary colouring matters

Carotenoids other than canthaxanthin:

Dissolve about 80 mg of canthaxanthin in 100 ml chloroform. Apply 400 µl of this solution as streak 2 cm from the bottom of the plate (Silicagel 0.25 mm).

Immediately develop the chromatogram with dichloromethane/ether (95+5) in a saturated chamber, suitably protected from light, until the solvent front has moved 15 cm above the initial streak. Remove the plate, allow the main part of the solvent to evaporate at room temperature and mark the principal band as well as the bands corresponding to other carotenoids. Remove the silica gel from the principal band, transfer it to a glass-stoppered 100-ml centrifuge tube and add 40.0 ml of a mixture of chloroform/methanol 3:1.
(Solution 1)

Separately remove the silica gel of the combined bands corresponding to other carotenoids, transfer them to a glass-stoppered, 50-ml centrifuge tube and add 20.0 ml of a mixture of chloroform/methanol 3:1.
(Solution 2)

Shake the centrifuge tubes by mechanical means for 10 minutes and centrifuge for 5 minutes. Dilute 10.0 ml of Solution 1 to 50.0 ml with chloroform.
(Solution 3)

Determine, with a suitable spectrophotometer, the absorbances of Solutions 2 and 3 in 1-cm cells at the wavelength of maximum in chloroform at about 485 nm, using chloroform as blank.

Calculation:

Carotenoids other than canthaxanthin (%) $= \dfrac{A_2 \cdot 10}{A_3}$

where A_2 = absorbance of Solution 2
A_3 = absorbance of Solution 3

*METHOD OF ASSAY   Weigh accurately about 0.1 g of the sample and proceed as directed in the spectrophotometric method in the Annex.

absorptivity (a) = 2200

Approximate wavelength of maximum absorption = 470 nm.

---

*   See Annex.

CAROTENES – NATURAL *
(Tentative)

SYNONYMS

DEFINITION — Carotenes are obtained by solvent** extraction of edible vegetables with subsequent removal of the solvent. The main colouring principle is Beta-carotene, but alpha- and gamma-carotenes may be present. Carotenes also contain other pigments and other substances such as oils, fats and waxes derived from the source material.

Class — Carotenoid

Code numbers — CI (1975) No. 75130
CI (1975) Food Orange 5
EEC Serial No. E160a

Chemical formula — $C_{40}H_{56}$

Structural formula — Beta-carotene has the following structure:

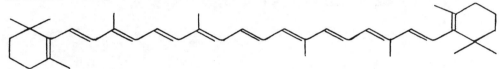

Molecular weight

Assay — Content not less than 0.5% of carotenes determined as Beta-carotene.

FUNCTIONAL USE — Food colour

CHARACTERISTICS

IDENTIFICATION TESTS

***A. Solubility — Insoluble in water

PURITY TESTS

Residual solvents — (Information required)
Arsenic — Not more than 3 mg/kg
Lead — Not more than 10 mg/kg
Heavy metals — Not more than 40 mg/kg

TESTS

METHOD OF ASSAY — (Information required)

---

* Supersedes the earlier specifications for Carotenes (Natural) published in FAO Nutrition Meetings Report Series (1959) Vol. II.
** Information required.
*** See General Methods (Guide to Specifications, FAO Food and Nutrition Paper No 5, Revision 1, 1983).

β-CAROTENE - SYNTHETIC*

SYNONYMS	CI Food Orange 5 β·β - Carotene
DEFINITION	
Class	Carotenoid
Code Numbers	CI (1975) No 40800 CI (1975) Food Orange 5 CAS No 7235-40-7 EEC No E 160a
Chemical names	β-carotene β·β - carotene
Chemical formula	$C_{40}H_{56}$

Structural formula

[Structural formula of β-carotene]

Molecular weight	536.44
Assay	Synthetic β-carotene is predominantly trans isomer and contains not less than 96% of total colouring matters.
DESCRIPTION	Red to brownish-red crystals or crystalline powder. β-carotene is sensitive to oxygen and light and should therefore be kept in a light-resistant container under inert gas. Users of β-carotene generally prefer to purchase the stabilized forms that are commercially available rather than the product that assays at not less than 96%. These stabilized forms are prepared from β-carotene, meeting the specification set out below, and include solutions or suspensions of β-carotene in edible fats or oils, emulsions and water dispersable powders. These preparations may contain permitted antioxidants, emulsifying agents and dispersing agents and may have different trans/cis isomer ratios from the parent colour described in the specification. The analytical methods prescribed for the parent colour are not necessarily suitable for the assay of, or determination of impurities in the stabilized forms.
FUNCTIONAL USE	Food colour

---

* Supersedes the earlier specifications for β-carotene-synthetic published in FAO Food and Nutrition Paper No 19 (1981).

CHARACTERISTICS

IDENTIFICATION TESTS

A. Solubility   Insoluble in water
Practically insoluble in methanol
Practically insoluble in ethanol
Slightly soluble in vegetable oils
Soluble in chloroform

B.   Determine the absorbance of the sample solution (see Method of Assay) at 455 nm and 483 nm. The ratio A_{455}/A_{483} is between 1.14 and 1.19.

C.   Determine the absorbance of the sample solution at 455 nm and that of sample Solution A (see Method of Assay) at 340 nm. The ratio A_{455}/A_{340} is not lower than 15.

D.   Passes test
(See description under TESTS)

E.   Passes test
(See description under TESTS)

PURITY TESTS

<u>Sulfated ash</u>   Not more than 0.1%
(See description under TESTS)

<u>Subsidiary colouring matters</u>

Carotenoids other than β-carotene:
Not more than 0.3%
(See description under TESTS)

**<u>Arsenic</u>   Not more than 3 mg/kg

**<u>Lead</u>   Not more than 10 mg/kg

*<u>Heavy metals</u>   Not more than 40 mg/kg

TESTS

IDENTIFICATION TESTS

D.   The colour of a solution of the sample in acetone disappears after the successive additions of a 5% solution of sodium nitrite and 1 <u>N</u> sulfuric acid.

E.   A solution of the sample in chloroform turns blue on addition of an excess of Carr-Price reagent TS.

---

* See General Methods (Guide to Specifications, FAO Food and Nutrition Paper No 5, Revision 1, 1983).

** See Annex.

PURITY TESTS

*Sulfated ash   Amount to weigh: 2.0 g

Subsidiary colouring matters

Carotenoids other than β-carotene:

Dissolve about 80 mg of β-carotene in 100 ml chloroform. Apply 400 μl of this solution as streak 2 cm from the bottom of the plate (silicagel 0.25 mm).

Immediately develop the chromatogram with cyclohexane/ether (80+20) in a saturated chamber suitably protected from light, until the solvent front has moved 15 cm above the initial streak.

Remove the plate. Allow the main part of the solvent to evaporate at room temperature and mark the principal band as well as the bands corresponding to other carotenoids. Remove the silica gel from the principal band, transfer it to a glass-stoppered 100-ml centrifuge tube and add 40.0 ml of chloroform. (Solution 1)

Separately remove the silica gel of the combined bands corresponding to other carotenoids, transfer to a glass-stoppered, 50-ml centrifuge tube and add 20.0 ml chloroform.
(Solution 2)

Shake the centrifuge tubes by mechanical means for 10 minutes and centrifuge for 5 minutes. Dilute 10.0 ml of Solution 1 to 50.0 ml with chloroform.
(Solution 3)

Determine, with a suitable spectrophotometer, the absorbances of Solutions 2 and 3 in 1 cm cells at the wavelength of maximum in chloroform at about 464 nm, using chloroform as blank.

Calculation:

Carotenoids other than β-carotene (%) = $\dfrac{A_2 \cdot 10}{A_3}$

where A_2 = absorbance of Solution 2; and
A_3 = absorbance of Solution 3.

**METHOD OF ASSAY

Weigh accurately about 0.08 g of the sample and proceed as directed in the spectrophotometric method in the Annex.

absorbtivity (a) = 2500

Approximate wavelength of maximum absorption = 455 nm.

---

* See General Methods (Guide to Specifications, FAO Food and Nutrition Paper No 5, Revision 1, 1983).

** See Annex.

CARTHAMUS RED*
(Tentative)

SYNONYMS	Carthamin Carthamic acid
DEFINITION	Carthamus Red is obtained from the dried petals of *Carthamus Tinctorius L*. The Carthamus Yellow is extracted from the petals with water and the residue treated with aqueous sodium hydroxide to extract the Carthamus Red. It is precipitated from the extract by addition of acid, separated by filtration and dried.
Code numbers	CI (1975) No 75140 CI (1975) Natural Red 26
Chemical name	Principally carthamin: 2-(β-D-glucopyranosyloxy)-3,5-dihydroxy-4-[1-hydroxy-3-(4-hydroxyphenyl)-2-propenylidene]-2,5-cyclohexadien-1-one
Chemical formula	$C_{21}H_{22}O_{10}$
Structural formula	Carthamin: (structure shown with $OC_6H_{11}O_5$ substituent)
Molecular weight	434.40
Assay	Content not less than 80% total colouring matters on a volatile matter-free basis.
DESCRIPTION	Dark red to red-brown powder with a characteristic slight odour.
FUNCTIONAL USE	Food colour

CHARACTERISTICS

IDENTIFICATION TESTS

**A. Solubility	Very slightly soluble in water Very slightly soluble in ethanol Practically insoluble in ether
B. Spectrophotometry	In dimethyl formamide λ max = 530 nm
C. Thin layer chromatograph	Rf = 0.54 - 0.55 (See description under TESTS)

---

* Supersedes the earlier specifications for Carthamus Red published in FAO Nutrition Meetings Report Series No 57 (1977).

** See General Methods (Guide to Specifications, FAO Food and Nutrition Paper No 5, Revision 1, 1983).

- 52 -

IDENTIFICATION TESTS (cont'd)

   D. Colour reaction   Passes test
                               (See description under TESTS)

   E. Colour reaction   Passes test
                               (See description under TESTS)

PURITY TESTS

*<u>Loss on drying at</u>   Not more than 5%
**o   (vacuo, over P_2O_5, 24 hours)

  <u>Synthetic dyes</u>   Absent
                              (See description under TESTS)

***<u>Arsenic</u>   Not more than 3 mg/kg

***<u>Lead</u>   Not more than 10 mg/kg

*<u>Heavy metals</u>   Not more than 40 mg/kg

TESTS

IDENTIFICATION TESTS

<u>C. Thin layer chromatography</u>

Activate some silica gel (Kiesel Gel G) for 1 hour at 100° and prepare a TLC plate. Prepare an 0.02% solution of the sample in methanol and apply 0.02 ml to the plate. Allow to dry and develop using a mixture of n-butanol, acetic acid and water (4:1:2 by volume) until the solvent front has ascended about 10 cm. Allow to dry and measure the Rf of the red spot.

<u>D. Colour reaction</u>

Dissolve 10 mg of the sample in 50 ml water. The colour of the solution is red. Add alkali to raise the pH to above 7. The colour changes to orange-yellow.

<u>E. Colour reaction</u>

To 0.05 g of the sample add 2 ml of 5% phosphoric acid, heat for 1 hour on a water bath. After cooling, filter and wash the residue with 3 ml of water. Combine the filtrate and the washings. Neutralize the combined solution with sodium hydroxide TS; add 5 ml of Fehling's TS and heat on a water bath for 10 minutes. A red precipitate is produced.

PURITY TESTS

<u>Synthetic dyes</u>

   (Method required)

---

*   See General Methods (Guide to Specifications, FAO Food and Nutrition Paper No 5, Revision 1, 1983).

**   Information required.

*** See Annex.

METHOD OF ASSAY

Place about 10 mg of the sample, previously dried, and accurately weighed, in 300-ml ground stoppered flask, add 150 ml of dimethylformamide, dissolve by shaking occasionally, and allow to stand for 2 hours. Filter this solution through a glass filter into a 200-ml volumetric flask. Wash the bottle and filter with two 25-ml portions of dimethylformamide, combine the filtrate and the washings, add dimethylformamide to volume, and mix. Determine the absorbance at 530 nm using 1-cm cells. Calculate the content using the absorptivity (*) for Carthamus Red.

---

*   Information required.

CARTHAMUS YELLOW*
(Tentative)

SYNONYM   Safflower yellow

DEFINITION   Carthamus Yellow is obtained by extracting the dried petals of *Carthamus Tinctorius L* with water and evaporating the extract to dryness.

    Code numbers   CI (1975) -
CI (1975) Natural Yellow 5
EEC Serial No. -

    Chemical formula $C_{21}H_{22}O_{11}$

    Structural formula

    Molecular weight   450.4

    Assay   (Information required)

DESCRIPTION   Yellowish brown to dark brown granular powder with green lustre

FUNCTIONAL USE   Food colour

CHARACTERISTICS

   IDENTIFICATION TESTS

    **A. Solubility   Soluble in water
Practically insoluble in ether

    **B. Melting point   About 230°

    C. Colour reaction   Passes test
(See description under TESTS

   PURITY TESTS

    Synthetic dyes   Absent
(See description under TESTS)

   ***Arsenic   Not more than 3 mg/kg

   ***Lead   Not more than 10 mg/kg

   **Heavy metals   Not more than 40 mg/kg

---

* Supersedes the earlier specifications for Carthamus Yellow published in FAO Nutrition Meetings Report Series No 57 (1977).

** See General Methods (Guide to Specifications, FAO Food and Nutrition Paper No 5, Revision 1, 1983).

*** See Annex.

TESTS

IDENTIFICATION TESTS

Colour reaction

Mix the sample with hydrochloric acid. A black colour appears, due to the formation of iso-carthamin.

PURITY TESTS

Synthetic dyes

Use paper chromatography. Carthamus Yellow has a very low Rf.

METHOD OF ASSAY

Determination of Total Colouring Matter by Spectrophotometry

Solvent: Citric acid buffer solution TS (pH 6.0)
Dilution of solution A: 100 ml → 250 ml
Absorptivity (a): 5.0
Approximate wavelength of maximum absorption: 400 nm

CHLOROPHYLLS*
(Tentative)

SYNONYMS

DEFINITION Chlorophylls are obtained by the solvent extraction of chlorophyll from grass, lucerne, nettle and other plants or from by-products of silk production with or without subsequent removal of magnesium from the complexes. Only the following solvents may be used:

    acetone, butanol, dichloromethane, ethanol, light petroleum, propan-2-ol, trichloroethylene and methyl ethyl ketone.

The solvent is subsequently removed. Chlorophylls also contain other pigments and other substances such as oils, fats and waxes derived from the source material.

Class Phorbin (=Dihydrophorphin)

Code numbers
CI (1975) No. 75810
CI (1975) Natural Green 3
CAS Numbers:
    Phaeophytin a, Magnesium complex: 479-61-8
    Phaeophytin b, Magnesium complex: 519-62-0
EEC No. E140

Chemical name The major colouring principles are phytyl 3-(4-ethyl-10-methoxycarbonyl-1,3,5,8-tetramethyl-9-oxo-2-vinylphorbin-7-yl) propionate (phaeophytin a) and phytyl 3-(4-ethyl-3-formyl-10-methoxycarbonyl-1,5,8-trimethyl-9-oxo-2-vinylphorbin-7-yl) propionate (phaeophytin b), present in the form of the magnesium complexes.

Chemical formula phaeophytin a, Magnesium complex: $C_{55}H_{72}MgN_4O_5$
phaeophytin b, Magnesium complex: $C_{55}H_{70}MgN_4O_6$

Structural formula The Magnesium complexes of phaeophytins a and b have the structure:

where X = CH_3 (phaeophyton a) or CHO (phaeophytin b)

---

* Supersedes the earlier specifications for Chlorophylls published in FAO Nutrition Meetings Report Series No. 46B (1969).

Molecular weight	Phaeophytin a, Magnesium complex: 893.50 Phaeophytin b, Magnesium complex: 907.49
Assay	Not less than 10% of total phaeophytins and complexes of phaeophytins, calculated as phaeophytin a.
DESCRIPTION	Waxy solids, ranging from olive green to dark green, depending on the magnesium content
FUNCTIONAL USE	Food colour

CHARACTERISTICS

IDENTIFICATION TESTS

*A. Solubility	Insoluble in water Soluble in ethanol Soluble in ether Soluble in chloroform Soluble in benzene
B. Colour reaction	Passes test (See description under TESTS)

PURITY TESTS

Residual solvents	Information required Acetone Butanol Dichloromethane Ethanol Light petroleum Propan-2-ol Trichloroethylene Methyl ethyl ketone
**Arsenic	Not more than 3 mg/kg
**Lead	Not more than 10 mg/kg
*Heavy metals	Not more than 40 mg/kg

TESTS

IDENTIFICATION TESTS

B. Colour reaction

Dissolve the sample in ether or petroleum ether. Add a small quantity of a 10% solution of potassium hydroxide in methanol. The colour changes to brown and then quickly returns to green.

METHOD OF ASSAY

(Method required)

---

* See General Methods (Guide to Specifications, FAO Food and Nutrition Paper No 5, Revision 1, 1983).

** See Annex.

CHLOROPHYLL-COPPER COMPLEX*
(Tentative)

SYNONYMS

DEFINITION Chlorophyll-copper complex is obtained by the solvent extraction of chlorophyll from grass, lucerne, nettle and other plants or from by-products of silk production with subsequent replacement of all or part of the magnesium by copper. Only the following solvents may be used:

actone, butanol, dichloromethane, ethanol, light petroleum, propan-2-ol, trichloro-ethylene and methyl ethyl ketone.

The solvent is subsequently removed.

Chlorophyll-copper complex also contains other pigments which may be present as copper derivatives and other substances such as oils, fats and waxes derived from the source material.

Class Porphyrin

Code numbers
CI (1975) No. 75810
CI (1975) Natural Green 3
CAS No.
EEC No. E141

Chemical name The major colouring principles are phytyl 3-(4-ethyl-10-methoxycarbonyl-1,3,5,8-tetramethyl-9-oxo-2-vinylphorbin-7-yl) propionate (phaeophytin a) and
phytyl 3-(4-ethyl-3-formyl-10-methoxycarbonyl-1,5,8-trimethyl-9-oxo-2-vinylphorbin-7-yl)propionate (phaeophytin b),
present in the form of the magnesium and/or copper complexes.

Chemical formula
Phaeophytin a, Copper complex: $C_{55}H_{72}CuN_4O_5$
Phaeophytin b, Copper complex: $C_{55}H_{70}CuN_4O_6$

Structural formula The copper complexes of Phaeophytins a and b have the structures:

where X = CH_3 (phaeophyton a) or CHO (phaeophytin b)

---

* Supersedes the earlier specifications for Chlorophyll-Copper Complex published in FAO Nutrition Meetings Report Series No 54B (1974).

Molecular weight	Phaeophytin a, Copper complex: 932.74 Phaeophyton b, Copper complex: 946.73
Assay	Not less than 10% of total phaeophytins and complexes of phaeophytins, calculated as phaeophytin a.

DESCRIPTION — Waxy solids, ranging from olive green to dark green depending on the magnesium and/or copper content.

FUNCTIONAL USE — Food colour

CHARACTERISTICS

IDENTIFICATION TESTS

*A. Solubility — Insoluble in water
Soluble in: ethanol, ether, chloroform, benzene

B. Colour reaction — Passes test
(See description under TESTS)

PURITY TESTS

<u>Residual solvents</u> — Information required

Acetone	Light petroleum
Butanol	Propan-2-ol
Dichloromethane	Trichloroethylene
Ethanol	Methyl ethyl ketone

<u>Free ionizable copper</u> — Not more than 200 mg/kg
(See description under TESTS)

**<u>Arsenic</u> — Not more than 3 mg/kg

**<u>Lead</u> — Not more than 10 mg/kg

*<u>Heavy metals</u> — Not more than 40 mg/kg

TESTS

IDENTIFICATION TESTS

<u>B. Colour reaction</u>

Dissolve the sample in ether or petroleum ether. Add a small quantity of a 10% solution of potassium hydroxide in methanol. The colour changes to brown and then quickly returns to green.

PURITY TESTS

<u>Free ionizable copper</u> — (Method required)

METHOD OF ASSAY — (Method required)

---

*   See General Methods (Guide to Specifications, FAO Food and Nutrition Paper No 5, Revision 1, 1983).
** See Annex.

CITRUS RED NO 2 *
(Tentative)

DEFINITION	Citrus Red No 2 consists essentially of 2-hydroxy-1-(2,5 dimethoxyphenylazo) naphthalene and subsidiary colouring matters
Class	Monoazo
Code numbers	CI (1975) No 12156 CI (1975) Solvent Red 80
Chemical name	2-hydroxy-1-(2,5-dimethoxyphenylazo) naphthalene
Chemical formula	$C_{18}H_{16}N_2O_3$
Structural formula	
Molecular weight	308.34
Assay	Content not less than 98% total colouring matter by spectrophotometry
DESCRIPTION	Orange-red powder
FUNCTIONAL USE	Food colour

CHARACTERISTICS

IDENTIFICATION TESTS

**A. Solubility — Insoluble in water
Soluble in aromatic hydrocarbons

***B. Identification of colouring matters

PURITY TESTS

***Loss on drying at 100°	Not more than 0.5%
Water soluble matter	Not more than 0.3% (See description under TESTS)

---

* Supersedes earlier specifications for Citrus Red No 2 published in FAO Nutrition Meetings Report Series No 46 (1969).

Information required on methods for determining subsidiary colouring matters and organic compounds other than colouring matters.

** See General Methods (Guide to Specifications, FAO Food and Nutrition Paper No 5, Revision 1, 1983).

*** See Annex.

PURITY TESTS (cont'd)

Subsidiary colouring matters	Not more than 2.0% (See description under TESTS)
Organic compounds other than colouring matters	
2,5-dimethoxyaniline 2-naphthol	Not more than 0.05% (See description under TESTS)
*Matter insoluble in carbon tetrachloride	Not more than 0.5%
*Arsenic	Not more than 3 mg/kg
*Lead	Not more than 10 mg/kg
**Heavy metals	Not more than 40 mg/kg

TESTS

PURITY TESTS

Water soluble matter

Apparatus

- Oven, 0-200° range
- Platinum dish
- Steam bath
- Desiccator

Procedure

Weigh accurately about 5 g (W_1) of a well powdered sample into a 500-ml conical flask fitted with a stopper. Add 200 ml distilled water. Stopper the flask and shake vigorously. Repeat this shaking several times during a 2 hour period. Filter. Add 100 ml of the filtrate to a weighed platinum dish (W_2).

Evaporate off the water on a steam bath. Dry the platinum dish and residue in an oven at 105°. Cool in a desiccator and then weigh the cooled dish (W_3).

Calculation

$$\% \text{ water soluble matter} = \frac{2(W_3 - W_2)}{W_1} \times 100$$

Subsidiary colouring matters

(Method required)

Organic compounds other than colouring matters

(Method required)

METHOD OF ASSAY

In the spectrophotometric method, use the following conditions:
    Solvent: Ethanol, 95%, reagent grade
    Dilution of solution A: 10 ml → 250 ml
    Absorptivity (a): 67.4
    Approximate wavelength of maximum absorption: 514 nm

---

*   See Annex.
** See General Methods (Guide to Specifications, FAO Food and Nutrition Paper No 5, Revision 1, 1983).

COCHINEAL AND CARMINIC ACID *

(Tentative)

DEFINITION	Cochineal extract is the concentrated solution obtained after removing the alcohol from an aqueous-alcoholic extract of cochineal *(Dactylopius coccus costa)*. The colouring principle is chiefly carminic acid. Commercial products also contain proteinaceous material derived from the source insect.
Class	Anthraquinone
Code numbers	CI (1975) No 75470 CI (1975) Natural Red 4 EEC Serial No E 120
Chemical name	7-D-glucopyranosyl-3,5,6,8-tetrahydroxy-1-methyl-9,10-dioxoanthracene-2-carboxylic acid
Chemical formula	$C_{22}H_{20}O_{13}$
Structural formula	

[Structural formula of carminic acid]

Molecular weight	492.40
Assay	Not less than 1.8% determined as carminic acid
DESCRIPTION	Dark red liquid
FUNCTIONAL USE	Food colour
CHARACTERISTICS	

IDENTIFICATION TESTS

**A. Solubility

B. Colour reaction — Passes test.
(See description under TESTS)

---

* Supersedes the earlier specification for Cochineal & Carminic Acid published in FAO Food and Nutrition Paper No 19 (1981).

** See General Methods (Guide to Specifications, FAO Food and Nutrition Paper No 5, Revision 1, 1983).

IDENTIFICATION TESTS (cont'd)

C. Colour reaction	Passes Test (See description under TESTS)
D. Colour reaction	Passes Test (See description under TESTS)
E. Colour reaction	Passes Test (See description under TESTS)

PURITY TESTS

Total solids	Not less than 5.7%, and not more than 6.3%. (See description under TESTS)
*Protein (non ammonia N x 6.25)	Not more than 2.2%
**Residual solvents Methyl alcohol	Not more than 150 mg/kg
**Arsenic	Not more than 3 mg/kg
**Lead	Not more than 10 mg/kg
***Heavy metals	Not more than 40 mg/kg
Salmonella	Absent (See description under TESTS)

TESTS

IDENTIFICATION TESTS

B. Colour reaction

Make the solution slightly alkaline by adding 1 drop of 10% sodium hydroxide or potassium hydroxide solution. A violet colour is produced.

C. Colour reaction

Add a small sodium dithionite ($Na_2S_2O_4$) crystal; the acid, neutral or alkaline solutions of Cochineal do not decolourize (difference from Orchil).

D. Colour reaction

Dry a small quantity of Cochineal in a porcelain dish. Cool thoroughly and treat the dry residue with 1 or 2 drops of cold sulfuric acid. No colour change occurs.

---

* See General Methods, Nitrogen Determination, Kjeldahl method (Guide to Specifications, FAO Food and Nutrition Paper No 5, Revision 1, 1983).

** See Annex.

*** See General Methods (Guide to Specifications, FAO Food and Nutrition Paper No 5, Revision 1 (1983).

TESTS (cont'd)

IDENTIFICATION TESTS (cont'd)

E. Colour reaction

Acidify a solution of Cochineal with 1/3rd its volume of hydrochloric acid and shake with amyl alcohol. Wash the amyl alcohol solution of Cochineal 2-4 times with an equal volume of water to remove hydrochloric acid, etc. Dilute amyl alcohol with 1-2 volumes of gasoline and shake with a few small portions of water to remove colour. Add, dropwise, 5% uranium acetate, shaking thoroughly after each addition. In the presence of Cochineal a characteristic emerald-green colour is produced.

PURITY TESTS

Total solids

(Information required)

Salmonella

Use the method given in the FDA Bacteriological Analytical Manual, Fifth Edition (1978), Chapter VI: Isolation and identification of salmonella.

METHOD OF ASSAY

Solvent: Water : 2N hydrochloric acid (97:3).

Weigh accurately about 800 mg of the sample and transfer with solvent to a 1000-ml volumetric flask. Dilute to volume with solvent and mix. Determine the absorbance of the solution in a 1 cm cell at the wavelength of maximum absorbance (about 494 nm) using solvent as the blank. If the measured absorbance of the solution is not within the range 0.20 to 0.25, then the weight of sample taken should be adjusted accordingly. Calculate the percentage of carminic acid in the sample taken for analysis by the formula:

$$(15) \cdot (A) \cdot (100) / (0.262) \cdot (X)$$

in which A = absorbance of the sample solution;
X = weight in mg of the sample taken; and
0.262 = absorbance of a solution of carminic acid having a concentration of 15 mg per litre.

CURCUMIN*

SYNONYMS	Curcumine Turmeric yellow Diferoylmethane
DEFINITION	Curcumin consists essentially of the pure colouring principle (1,7-bis(4-hydroxy-3-methoxyphenyl)hepta-1,6-diene-3,5-dione) obtained by solvent extraction of turmeric, which consists of the ground rhizomes of *Curcuma longa L.* Only the following solvents may be used in the extraction: acetone, dichloromethane, ethylene dichloride, methanol, ethanol, light petroleum.
Code numbers	CI (1975) No. 75300 CI (1975) Natural Yellow 3 EEC Serial No. E100
Chemical name	1,7-bis-(4-hydroxy-3-methoxyphenyl)-1,6-heptadiene-3,5-dione
Chemical formula	$C_{21}H_{20}O_6$
Structural formula	
Molecular weight	368.37
Assay	Content not less than 90% total colouring matters
DESCRIPTION	Orange-yellow crystalline powder
FUNCTIONAL USE	Food colour

CHARACTERISTICS

IDENTIFICATION TESTS

**A.	Solubility	Insoluble in water and in ether. Soluble in ethanol and in glacial acetic acid.
**B.	Melting range	179-182°
C.	Colour reaction	Passes test (See description under TESTS)

---

* Supersedes the earlier specifications for Curcumin published in FAO Food and Nutrition Paper No 25 (1982).
** See General Methods (Guide to Specifications, FAO Food and Nutrition Paper No 5, Revision 1, 1983).

IDENTIFICATION TESTS (cont'd)

   D. Colour reaction   Passes test
                            (See description under TESTS)

   E. Chromatography   Passes test
                            (See description under TESTS)

PURITY TESTS

*__Residual solvents__	Acetone Dichloromethane Ethylene dichloride Methanol Ethanol Light petroleum	Not more than 50 mg/kg, singly or in combination.
*__Arsenic__	Not more than 3 mg/kg	
*__Lead__	Not more than 10 mg/kg	
**__Heavy metals__	Not more than 40 mg/kg	

TESTS

IDENTIFICATION TESTS

  C. Colour reaction

    A solution of the sample in ethanol is characterized by pure yellow colour and light green fluorescence; if this ethanol extract is added to concentrate sulfuric acid, a deep crimson colour is produced.

  D. Colour reaction

    Treat an aqueous or dilute ethanolic solution of the sample with hydrochloric acid until a slightly orange colour begins to appear. Divide mixture into 2 parts and add some boric acid powder or crystals to one portion. Marked reddening will be quickly apparent, best seen by comparison with portion to which the boric acid has not been added. The test may also be made by dipping pieces of filter paper in ethanolic solution of colouring matter, drying at 100°, and then moistening with a weak solution of boric acid to which a few drops of hydrochloric acid have been added. On drying, a cherry-red colour will develop.

  E. Chromatography

    Spot 5 μl of test solution 0.02 g p. 100 μg in ethanol 95% on a TLC plate of silica gel GF254. Place the plate in a developing chamber containing the mixture dichloromethane/isopropylether/methanol (1:8.5:0.5) as solvent and allow the solvent front to ascend about 15 cm. Examine under daylight and under U.V. light:
    - one spot Rf about 0.4 under yellow fluorescence and daylight, and
    - two other spots Rf about 0.5 under U.V. light, visible.

---

*   See Annex.

**   See General Methods (Guide to Specifications, FAO Food and Nutrition Paper No 5, Revision 1, 1983).

METHOD OF ASSAY

Procedure

Place on a water bath at 90° a 100-ml graduated flask containing 60 ml of glacial acetic acid and 0.10 g of the sample. Keep it on the bath for 1 hour. Then add 2 g of boric acid and 2 g of oxalic acid and keep the flask on the bath for 10 minutes more. Cool to room temperature and fill it to the mark with glacial acetic acid. Shake very well. Pipet 5 ml of the mixture into a 50-ml measuring flask and fill it to the mark with glacial acetic acid. Then measure the absorbance of the red coloured solution in a 1-cm cell at 540 nm using glacial acetic acid as the reference.

Calibration Curve

Dissolve 0.10 g of pure curcumin, by heating, with glacial acetic acid in a 100-ml measuring flask. Pipet 5,10,15 ml, etc. of this solution into 100-ml measuring flasks together with boric acid and oxalic acid as under Procedure, above. Prepare the calibration curve and determine the curcumin concentration corresponding to the absorbance of the sample solution.

ERYTHROSINE *

SYNONYMS	CI Food Red 14 FD&C Red No 3
DEFINITION	Erythrosine consists essentially of disodium 2-(2,4,5,7-tetraiodo-3-oxido-6-oxoxanthen-9-yl) benzoate and subsidiary colouring matters together with sodium chloride and/or sodium sulfate as the principal uncoloured components. Erythrosine may be converted to the corresponding aluminium lake in which case only the General Specifications for Aluminium Lakes of Colouring Matters shall apply.
Class	Xanthene
Code numbers	CI (1975) No 45430 CI (1975) Food Red 14 CAS No 16423-68-0 EEC No E 127
Chemical name	Disodium 2-(2,4,5,7-tetraiodo-3-oxido-6-oxoxanthen-9-yl) benzoate
Chemical formula	$C_{20}H_6I_4O_5Na_2$
Structural formula	
Molecular weight	879.87
Assay	Content not less than 85% total colouring matters
DESCRIPTION	Red powder or granules
FUNCTIONAL USE	Food colour

CHARACTERISTICS

IDENTIFICATION TESTS

**A. Solubility	Soluble in water Soluble in ethanol

---

* Supersedes the earlier specification for Erythrosine published in FAO Food and Nutrition Paper No 19 (1981).

** See General Methods (Guide to Specifications, FAO Food and Nutrition Paper No 5, Revision 1, 1983).

CHARACTERISTICS (cont'd)

IDENTIFICATION TESTS (cont'd)

*B. Identification of colouring matters

PURITY TESTS

*Loss on drying at 135°	Not more than 15%
*Chloride and sulfate calculated as sodium salts	Not more than 15%
Inorganic iodides calculated as sodium iodide	Not more than 0.1% (See description under TESTS)
*Water insoluble matter	Not more than 0.2%
Subsidiary colouring matters (except Fluorescein)	Not more than 4% (See description under TESTS)
Fluorescein	Not more than 20 mg/kg (See description under TESTS)

Organic compounds other than colouring matters

Tri-iodo resorcinol:	Not more than 0.2%	(See description under TESTS)
2-(2,4-dihydroxy-3,5-di-iodobenzoyl) benzoic acid:	Not more than 0.2%	
*Ether extractable matter	From a solution of pH, not less than 7, not more than 0.2%	
*Arsenic	Not more than 3 mg/kg	
*Lead	Not more than 10 mg/kg	
*Zinc	Not more than 50 mg/kg	
**Heavy metals	Not more than 40 mg/kg	

TESTS

PURITY TESTS

Inorganic iodides calculated as sodium iodide

Principle
The sodium iodide content of Erythrosine is determined by potentiometric titration with silver nitrate solution using an iodide ion selective electrode.

Apparatus
- Iodide specific electrode with suitable millivoltmeter and reference electrode.
- Magnetic stirrer.

---

*   See Annex.
**   See General Methods (Guide to Specifications, FAO Food and Nutrition Paper No 5, Revision 1, 1983).

TESTS (cont'd)

Inorganic iodides calculated as sodium iodide (cont'd)

Reagents
0.001 M silver nitrate solution

Procedure
Weigh 1.0 g of the sample into a 100-ml beaker. Add 75 ml distilled water and the magnetic follower. Stir to dissolve.

Immerse the iodide and reference electrodes in the solution and set the meter to read the potential of the system in millivolts.

Add silver nitrate solution from a burette initially in 0.5 ml aliquots, reducing these to 0.1 ml as the end-point approaches indicated by an increasing change in potential for each addition. After allowing time for the reading to stabilize, record the millivolt readings after each addition.

Continue the titration until further additions make little change in the potential.

Plot the millivolt readings against the volume of silver nitrate solution added. The equivalent point is the volume corresponding to the maximum slope of the curve.

Calculation
Percentage sodium iodide in sample = Titre x 0.015%.

Subsidiary colouring matters

Use the following conditions:
Developing solvent: No 5
Height of ascent of solvent front: 17 cm

Note: Take special care not to allow the chromatograms to be exposed to direct sunlight.

Fluorescein

Principle
The fluorescein is separated from the Erythrosine by TLC and compared with a standard chromatogram prepared from fluorescein at the concentration corresponding to the limit figure.

Solvents
Methanol   500 ml
Water   400 ml
Ammonia (s.g. 0.890) 100 ml

Sample
Weigh 1 g sample, dissolve in about 50 ml solvent and dilute to 100 ml in a volumetric flask.

./.

TESTS (cont'd)

Fluorescein (cont'd)

Standard
Weigh an amount of the fluorescein standard corresponding to 1 g at the colouring matter content of the Erythrosine sample. Dissolve in water (or in water with 10 ml ammonia s.g. 0.890 if fluorescein free acid is being used) and dilute to 100 ml. Make further sequential dilutions as follows:

  1 ml to 100 ml with water
  1 ml to 100 ml with water
  20 ml to 100 ml with solvent

Chromatography solvent

n-butanol	100 ml
water	44 ml
ammonia (s.g. 0.890)	1 ml
ethanol	22.5 ml

TLC
Spot 25 µl of the sample and standard solutions side by side on a cellulose plate. Develop for 16 hours in the chromatography solvent. Allow the plate to dry. View under a UV light source and compare the fluorescence of the standard with the fluorescence of the corresponding area on the chromatogram of the sample. The intensity of the latter shall not be greater than that of the former.

Note: Take special care not to allow the chromatograms to be exposed to direct sunlight.

Organic compounds other than colouring matters

Proceed as directed under Column Chromatography.

The following absorptivities may be used:

2(2,4-dihydroxy-3,5-di-iodobenzoyl) benzoic acid:
    0.047 mg/L/cm at 348 nm (alkaline).

Tri-iodoresoreinol: 0.079 mg/L/cm at 223 nm (acidic).

METHOD OF ASSAY

Dissolve about 1.00 g of the sample, accurately weighed, in 250 ml of water, transfer to a clean 500-ml beaker, add 8.0 ml of 1.5 N nitric acid and stir well. Filter through a sintered glass crucible (porosity 3, diameter 5 cm) which has been weighed containing a small glass stirring rod. Wash thoroughly with 0.5% nitric acid until the filtrate gives no turbidity with silver nitrate TS, and then wash with 30 ml water. Dry to constant weight at $135° \pm 5°$, carefully breaking up the precipitate by means of the glass rod. Cool in desiccator and weigh.

$$\text{Percent total colouring matters} = \frac{\text{Weight of residue} \times 105.3}{\text{Weight of sample}}$$

Determination of Hydrochloric Acid-insoluble Matter in Erythrosine Lake

Reagents

    Concentrated hydrochloric acid
    Hydrochloric acid 0.5% v/v
    Dilute ammonia solution (dilute 10 ml ammonia s.g. 0.890 to 100 ml with water).

Procedure

Accurately weigh approximately 5 g of the lake into a 500-ml beaker. Add 250 ml water and 60 ml concentrated hydrochloric acid. Boil to dissolve the alumina while the Erythrosine converts to its "free acid" form, which is insoluble in acid. Filter through a tared No 4 sintered glass crucible. Wash the crucible with a small amount of hot 0.5% hydrochloric acid and then with some hot distilled water. Remove the acid filtrate from the filter flask, replace the crucible and wash with hot dilute ammonia solution until the washings are colourless. Dry the crucible to constant weight at 135°. Express the residue as a percentage of the weight taken.

FAST GREEN FCF *
(Tentative)

SYNONYMS	CI Food Green 3 FD&C Green 3
DEFINITION	Fast Green FCF consists essentially of inner salt disodium salt of N-ethyl-N-[4-[[4-[ethyl [(3-sulfophenyl) methyl] amino] phenyl](4-hydroxy-2-sulfophenyl) methylene]-2,5-cyclohexadien-1-ylidene]-3-sulfobenzenemethanaminium hydroxide and isomers and subsidiary colouring matters together with sodium chloride and/or sodium sulfate as the principal uncoloured components. Fast Green FCF may be converted to the corresponding aluminium lake in which case only the General Specifications for Aluminium Lakes of Colouring Matters shall apply.
Class	Triarylmethane
Code numbers	CI (1975) No 42053 CI (1975) Food Green 3 CAS No 2353-43-9
Chemical name	Inner salt disodium salt of N-ethyl-N-[4-[[4-[ethyl [(3-sulfophenyl) methyl] amino] phenyl](4-hydroxy-2-sulfophenyl) methylene]-2,5-cyclohexadien-1-ylidene]-3-sulfobenzenemethanaminium hydroxide
Chemical formula	$C_{37}H_{34}N_2Na_2O_{10}S_3$
Structural formula	
Molecular weight	808.84
Assay	Content not less than 85% total colouring matters
DESCRIPTION	Red to brown-violet powder or crystals
FUNCTIONAL USE	Food colour

---

* Supersedes the earlier specification for Fast Green FCF published in FAO Nutrition Meeting's Report Series No 38B (1966).

Information required on the method of analysis for subsidiary colouring matters.

CHARACTERISTICS

IDENTIFICATION TESTS

*A. Solubility   Soluble in water
                                             Slightly soluble in ethanol

**B. Identification of colouring matters

PURITY TESTS

**Loss on drying at 135°	Not more than 15%
**Chloride and sulfate calculated as sodium salts	Not more than 15%
**Water insoluble matter	Not more than 0.2%
Subsidiary colouring matters	Not more than 6% (See description under TESTS)

Organic compounds other than colouring matters

    Sum of 2-, 3-, and 4-formylbenzene sulfonic acids, sodium salts:
        Not more than 0.5%

    Sum of 3- and 4-[[ethyl](4-sulfophenyl)amino]methyl benzene
       sulfonic acid, disodium salts:
        Not more than 0.3%

    2-formyl-5-hydroxybenzenesulfonic acid, sodium salt:
        Not more than 0.5%

    Leuco base: Not more than 5.0%

    **Unsulfonated primary aromatic amines:
        Not more than 0.01% calculated as aniline

**Ether extractable matter	Not more than 0.4%
**Arsenic	Not more than 3 mg/kg
**Lead	Not more than 10 mg/kg
**Chromium	Not more than 50 mg/kg
*Heavy metals	Not more than 40 mg/kg

---

*   See General Methods (Guide to Specifications, FAO Food and Nutrition Paper No 5, Revision 1, 1983).

** See Annex.

TESTS

PURITY TESTS

*Subsidiary colouring matters

   Use the following conditions:
   Developing solvent: (Information required)
   Height of ascent of solvent front:

*Organic compounds other than colouring matters

   Proceed as directed under Column Chromatography.

   The following absorptivities may be used:

   3-formyl benzene sulfonic acid:
       0.495 mg/L/cm at 246 nm in dilute HCl

   3-[(ethyl)(4-sulfophenyl)amino] methyl benzene sulfonic acid:
       0.078 mg/L/cm at 277 nm in dilute ammonia

   2-formyl-5-hydroxybenzene sulfonic acid:
       0.080 mg/L/cn at 335 nm in dilute ammonia

*Leuco base

   Weigh accurately 130 ± 5 mg sample and proceed as directed under the method for the determination of Leuco base.

   Absorptivity (a) = 0.156/mg/L/cm at approx. 625 nm
   Ratio = 0.9712

METHOD OF ASSAY

*Determination of Total Colouring Matters by Titration with Titanous Chloride

   Use the following:
   Weight of sample: 1.9 - 2.0 g
   Buffer: 15 g sodium hydrogen tartrate
   Weight (D) of colouring matters equivalent to
       1.00 ml of 0.1 N $TiCl_3$: 0.04045 g

---

*   See Annex.

FAST RED E *

SYNONYM	CI Food Red 4
DEFINITION	Fast Red E consists essentially of disodium 2-hydroxy-1-(4-sulfonato-1-naphthylazo) naphthalene-6-sulfonate and subsidiary colouring matters together with sodium chloride and/or sodium sulfate as the principal uncoloured components.
	Fast Red E may be converted to the corresponding aluminium lake in which case only the General Specifications for Aluminium Lakes of Colouring Matters shall apply.
Class	Monoazo
Code numbers	CI (1975) No 16045 CI (1975) Food Red 4 CAS No 2302-96-7
Chemical name	Disodium 2-hydroxy-1-(4-sulfonato-1-naphthylazo) napthalene-6-sulfonate
Chemical formula	$C_{20}H_{12}N_2Na_2O_7S_2$
Structural formula	
Molecular weight	502.44
Assay	Content not less than 85% total colouring matters
DESCRIPTION	Red-brown powder or granules
FUNCTIONAL USE	Food colour
CHARACTERISTICS	

IDENTIFICATION TESTS

**A. Solubility Soluble in water
Sparingly soluble in ethanol

***B. Identification of colouring matters

---

* Supersedes the earlier specifications for Fast Red E published in FAO Nutrition Meetings Report Series No 57 (1977).

** See General Methods (Guide to Specifications, FAO Food and Nutrition Paper No 5, Revision 1, 1983).

*** See Annex.

CHARACTERISTICS (cont'd)

PURITY TESTS

*Loss on drying at 135°	Not more than 15%
*Chloride and sulfate calculated as sodium salts	Not more than 15%
*Water insoluble matter	Not more than 0.2%
Subsidiary colouring matters	Not more than 3%

Organic compounds other than colouring matters

    4-aminonaphthalene-1-sulfonic acid
    6-hydroxynaphthalene-2-sulfonic acid
    6,6'-oxydi(naphthalene-2-sulfonic acid)

Total not more than 0.5% (See description under TESTS)

*Unsulfonated primary aromatic amines:
    Not more than 0.01% calculated as aniline

*Ether extractable matter	Not more than 0.2%
*Arsenic	Not more than 3 mg/kg
*Lead	Not more than 10 mg/kg
**Heavy metals	Not more than 40 mg/kg

TESTS

PURITY TESTS

*Subsidiary colouring matters

    Use the following conditions:

        Developing solvent: No 3
        Height of ascent of solvent front: approximately 12 cm

    Proceed as directed under Column Chromatography.

*Organic compounds other than colouring matters

    Proceed as directed under Column Chromatography.

METHOD OF ASSAY

*Determination of Total Colouring Matters by Titration with Titanous Chloride

    Use the following:

        Weight of sample: 0.5 - 0.6 g
        Buffer: 15 g sodium hydrogen tartrate
        Weight (D) of colouring matters equivalent to
            1.00 ml of 0.1 N $TiCl_3$: 0.01256 g

---

*   See Annex.

** See General Methods (Guide to Specifications, FAO Food and Nutrition Paper No 5, Revision 1, 1983).

FERROUS GLUCONATE*

SYNONYMS

DEFINITION

    Class

    Code number   CI (1975) -
                              CI (1975) -

    Chemical name   Ferrous gluconate
                              Iron (II) di-D-gluconate dihydrate

    Chemical formula $C_{12}H_{22}FeO_{14} \cdot 2H_2O$

    Structural formula

$$\left(\begin{array}{c} COO- \\ HCOH \\ HOCH \\ HCOH \\ HCOH \\ CH_2OH \end{array} \right)_2 Fe \cdot 2H_2O$$

    Molecular weight   482.17

    Assay   Content not less than 95% of $C_{12}H_{22}FeO_{14}$

DESCRIPTION   Fine yellowish-grey or pale greenish-yellow powder or granules having a slight odour resembling that of burnt sugar.

FUNCTIONAL USE   Colouring adjunct

CHARACTERISTICS

    IDENTIFICATION TESTS

    **A. Solubility   Soluble with slight heating in water
                            Practically insoluble in ethanol

    B. Test for Fe II   Passes test

    C. Chemical reaction   Passes test

---

*   Supersedes the earlier specifications for Ferrous Gluconate published in FAO Nutrition Meetings Report Series No 54B (1974).

**   See General Methods (Guide to Specifications, FAO Food and Nutrition Paper No 5, Revision 1, 1983).

PURITY TESTS

*Loss on drying at 105°	Not less than 6.5% after 4 hours. Not more than 10% after 4 hours.
Other organic compounds	
Oxalic acid	Passes test (See description under TESTS)
Reducing sugars	Passes test (See description under TESTS)
**Arsenic	Not more than 3 mg/kg
**Lead	Not more than 10 mg/kg
**Mercury	Not more than 3 mg/kg
Iron (Fe III)	Not more than 2% (See description under TESTS)

TESTS

IDENTIFICATION TESTS

C. Chemical reaction

To 5 ml of a warm 1-in-10 solution add 0.65 ml of glacial acetic acid and 1 ml of freshly distilled phenylhydrazine. Heat the mixture on a steam bath for 30 minutes. Cool, and scratch the inner surface of the container with a glass stirring rod. Crystals of gluconic acid phenylhydrazide form.

PURITY TESTS

Oxalic acid

Dissolve 1 g of the sample in 10 ml of water and add 2 ml of hydrochloric acid. Using a separatory funnel, extract successively with 50 and 20 ml of ether. Combine the ether extracts, add 10 ml of water, and evaporate the ether on a steam bath. Add 1 drop of 36% acetic acid and 1 ml of a 1-in-20 calcium acetate solution. No turbidity should be produced within 5 minutes.

Reducing sugars

Dissolve 0.5 g of the sample in 10 ml of water; warm, and make the solution alkaline with 1 ml of ammonia TS. Pass hydrogen sulfide gas into the solution to precipitate the iron, and allow the mixture to stand for 30 minutes to coagulate the precipitate. Filter, and wash the precipitate with two successive 5-ml portions of water. Acidify the combined filtrate and washings with hydrochloric acid, and add 2 ml of dilute hydrochloric acid TS in excess. Boil the solution until the vapours no longer darken lead acetate paper, and continue to boil, if necessary, until concentrated to about 10 ml. Allow to cool, add 5 ml of sodium carbonate TS and 20 ml of water; filter, and adjust the volume of the filtrate to 100 ml. To 5 ml of the filtrate add 2 ml of alkaline cupric tartrate TS and boil for 1 minute. No red precipitate should be formed within 1 minute.

---

* See General Methods (Guide to Specifications, FAO Food and Nutrition Paper No 5, Revision 1, 1983).

** See Annex.

PURITY TESTS (cont'd)

Iron (Fe III)

Dissolve about 5 g of the sample, accurately weighed, in a mixture of 100 ml of water and 10 ml of hydrochloric acid in a 250-ml glass stoppered flask. Add 3 g of potassium iodide, shake well, and allow to stand in the dark for 5 minutes. Titrate any liberated iodine with 0.1 N sodium thiosulfate, using starch TS as the indicator. Each ml of 0.1 $\underline{N}$ sodium thiosulfate is equivalent to 5.585 mg of iron (Fe III).

METHOD OF ASSAY

Dissolve about 1.5 g of previously dried sample, accurately weighed, in a mixture of 75 ml of water and 15 ml of dilute sulfuric acid TS in a 300-ml Erlenmeyer flask, and add 250 mg of zinc dust.

Close the flask with a stopper containing a Bunsen valve, and allow to stand at room temperature for 20 minutes. Then filter through a Gooch crucible containing a glass fibre filter paper coated with a thin layer of zinc dust, and wash the crucible and contents with 10 ml of dilute sulfuric acid TS, followed by 10 ml of water.

Add orthophenanthroline TS and titrate the filtrate in the suction flask immediately with 0.1 $\underline{N}$ ceric sulfate. Perform a blank determination, and make any necessary correction.

Each ml of 0.1 $\underline{N}$ ceric sulfate is equivalent to 44.61 mg of $C_{12}H_{22}FeO_{14}$.

GRAPE SKIN EXTRACT *

SYNONYMS	Enociania; Eno
DEFINITION	Grape skin extract is obtained by aqueous extraction of grape skin or marc after the juice has been expressed from it. It contains the common components of grape juice, namely: anthocyanine, tartaric acid, tannins, sugars, minerals, etc., but not in the same proportions as found in grape juice. During the extraction process, sulphur dioxide is added and most of the extracted sugars are fermented to alcohol. The extract is concentrated by vacuum evaporation during which practically all the alcohol is removed. A small amount of sulphur dioxide may be present.
Class	Anthocyanin
Code numbers	CI (1975) - none EEC Serial No E 163
Chemical name	The principal colouring matters are anthocyanins glucosides of anthocyanidins (2-phenylbenzopyrylium salts) such as peonidin, malvidin, delphinidin, and petunidin.
Chemical formulae	Peonidin: $C_{16}H_{13}O_6X$ Malvidin: $C_{17}H_{15}O_7X$ Delphinidin: $C_{15}H_{11}O_7X$ Petunidin: $C_{16}H_{13}O_7X$ X: acid moiety
Structural formula	[structure of anthocyanidin cation] Peonidin: R = OCH₃; R' = H Malvidin: R, R' = OCH₃ Delphinidin: R, R' = OH Petunidin: R = OCH₃; R' = OH X⁻: acid moiety
Molecular weight	
Assay	The colour intensity is not less than declared.

---

* Supersedes the earlier specifications for Grape Skin Extract published in FAO Food and Nutrition Paper No 25 (1982).

DESCRIPTION   Purplish-red liquid, lump, powder or
                                          paste, having a slight characteristic
                                          odour.

FUNCTIONAL USE   Food colour

CHARACTERISTICS

  IDENTIFICATION TESTS

  A. Solubility   Soluble in water

  B. Spectrophotometry   At pH_3 the absorbance maximum is
                                          about 525 nm

  C. Colour reaction   Passes test
                                          (See description under TESTS)

  PURITY TESTS

  Basic colouring matters   Passes test
                                          (See description under TESTS)

  Other acidic colouring matters   Passes test
                                          (See description under TESTS)

  Sulfur dioxide   Not more than 0.005% per 1 colour value
                                          (See description under TESTS)

  *Arsenic   Not more than 3 mg/kg

  *Lead   Not more than 10 mg/kg

  **Heavy metals   Not more than 40 mg/kg

TESTS

  IDENTIFICATION TESTS

  C. Colour reaction

   Add 0.1 g of the sample to 50 ml of water and shake thoroughly.
   Filter if necessary. The solution shows red to purplish-red
   colour and it turns to blue or dark green on the addition of
   sodium hydroxide TS.

  PURITY TESTS

  Basic colouring matters

   Add 1 g of the sample to 100 ml sodium hydroxide solution (1 in
   100) and shake well. Take 30 ml of this solution and extract with
   15 ml of ether. Extract this ether extract twice with each 5 ml
   of dilute acetic acid TS. The acetic acid extract is colourless.

---

*   See Annex.
** See General Methods (Guide to Specifications, FAO Food and Nutrition
     Paper No 5, Revision 1 (1983).

TESTS (cont'd)

PURITY TESTS (cont'd)

Other acidic colouring matters

Add 1 ml of ammonia TS and 10 ml of water to 1 g of the sample and shake well. Following the directions in the General Methods for Chromatography* place 0.002 ml of the solution on the chromatographic sheet and dry it. Use a mixture of pyridine and ammonia TS (2:1 by volume) as developing solvent and stop the development when the solvent front reaches about 15 cm height from the point where the sample solution was placed. No spot is observed at the solvent front after drying under daylight. If any spot is observed, it should be decolourized when sprayed with a solution of stannous chloride in hydrochloric acid (2 in 5).

Sulphur dioxide

Distil 1 g of the sample with 100 ml of water and 25 ml of phosphoric acid solution (2 in 7) in a distilling flask with the Wagner tube (Fig. 1). In an absorption flask, place 25 ml of lead acetate solution (1 in 50) previously prepared. Insert the lower end of condenser into the lead acetate solution in the absorption flask. Distil until the liquid in the absorption flask reaches about 100 ml and rinse the end of condenser with a little amount of water. To the distilled solution add 5 ml of hydrochloric acid and 1 ml of starch TS, and titrate with 0.01 N iodine. Each ml of 0.01 N iodine is equivalent to 0.3203 mg of SO_2.

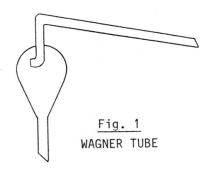

Fig. 1
WAGNER TUBE

METHOD OF ASSAY

In the absence of an assay method, a measurement of colour intensity by the following method may be used.

Prepare approximately 200 ml of pH 3.0 citric acid - dibasic sodium phosphate buffer solution:

Mix 159 volumes of 2.1% citric acid solution and 41 volumes of 0.16% dibasic sodium phosphate solution, and adjust the pH to 3.0, using the citric acid solution or dibasic sodium phosphate solution.

Weigh accurately an adequate amount of the sample so that the measured absorbance is between 0.2 and 0.7, and add pH 3.0 citric acid - dibasic sodium phosphate buffer solution to make up a 100-ml solution. Measure the absorbance A of this solution in a 1 cm cell at the wavelength of maximum absorption around 525 nm, using pH 3.0 citric acid - dibasic sodium phosphate buffer solution as the blank.

$$\text{Colour value} = \frac{A \times 10}{\text{Weight of sample (g)}}$$

---

* See General Methods for Chromatography, B. Paper Chromatography (3) Procedure for Ascending Chromatography (Guide to Specifications, FAO Food and Nutrition Paper No 5, Revision 1, 1983).

GREEN S *

SYNONYM	CI Food Green 4
DEFINITION	Green S consists essentially of sodium 5-[4-dimethylamino-α-(4-dimethyliminiocyclohexa-2,5-dienylidene) benzyl]-6-hydroxy-7-sulfonatonaphthalene-2-sulfonate and subsidiary colouring matters together with sodium chloride and/or sodium sulfate as the principal uncoloured components.
	Green S may be converted to the corresponding aluminium lake, in which case only the General Specifications for Aluminium Lakes of Colouring Matters shall apply.
Class	Triarylmethane
Code numbers	CI (1975) No 44090 CI (1975) Food Green 4 CAS No 860-22-0 EEC No E 142
Chemical name	Sodium 5-[4-dimethylamino-α-(4-dimethyliminiocyclohexa-2,5-dienylidene) benzyl]-6-hydroxy-7-sulfonatonaphthalene-2-sulfonate.
Chemical formula	$C_{27}H_{25}N_2Na\,O_7S_2$
Structural formula	

Molecular weight	576.63
Assay	Content not less than 80% total colouring matters
DESCRIPTION	Dark green powder or granules
FUNCTIONAL USE	Food colour
CHARACTERISTICS	

IDENTIFICATION TESTS

**A. Solubility	Soluble in water Slightly soluble in ethanol

---

*   Supersedes the earlier specification for Food Green S published in FAO Nutrition Meeting's Report Series No 54B (1975).

** See General Methods (Guide to Specifications, FAO Food and Nutrition Paper No 5, Revision 1, 1983).

IDENTIFICATION TESTS (cont'd)

*B. Identification of colouring matters

PURITY TESTS

*<u>Loss on drying at 135°</u>	Not more than 20%
*<u>Chloride and sulfate calculated as sodium salts</u>	Not more than 20%
*<u>Water insoluble matter</u>	Not more than 0.2%
<u>Subsidiary colouring matters</u>	Not more than 1% (See description under TESTS)

<u>Organic compounds other than colouring matters</u>

4,4'-bis (dimethylamino) benzhydryl alcohol:	Not more than 0.1%	
4,4'-bis (dimethylamino) benzophenone:	Not more than 0.1%	(See description under TESTS)
3-hydroxynaphthalene-2,7-disulfonic acid:	Not more than 0.2%	
Leuco base:	Not more than 5.0%	

*Unsulfonated primary aromatic amines:	Not more than 0.01% calculated as aniline
*<u>Ether extractable matter</u>	Not more than 0.2%
*<u>Arsenic</u>	Not more than 3 mg/kg
*<u>Lead</u>	Not more than 10 mg/kg
*<u>Chromium</u>	Not more than 50 mg/kg
**<u>Heavy metals</u>	Not more than 40 mg/kg

TESTS

*<u>Subsidiary colouring matters</u>

Use the following conditions:
Developing solvent: No. 2
Height of ascent of solvent front: approximately 17 cm

*<u>Organic compounds other than colouring matters</u>

HPLC elution gradient: 2 to 100% at 2.0% per minute (linear)

*<u>Leuco base</u>

Weigh accurately 110 ± 5 mg of sample and proceed as directed under the Method for the Determination of Leuco Base.

Absorptivity: (a) = 0.1725/mg/L/cm at approximately 634 nm
Ratio = 0.9600

---

* See Annex.

** See General Methods (Guide to Specifications, FAO Food and Nutrition Paper No 5, Revision 1, 1983).

METHOD OF ASSAY

Determination of Total Colouring Matters by Titration with Titanous Chloride

Use the following conditions:
   Weight of sample: 1.4-1.5 g
   Buffer: 15 g sodium hydrogen tartrate
   Weight (D) of colouring matters equivalent to 1.00 ml of
      0.1 N $TiCl_3$: 0.02883 g

---

*   See Annex.

INDIGOTINE*

SYNONYMS	CI Food Blue 1 FD&C Blue No.2 Indigo Carmine
DEFINITION	Indigotine consists essentially of a mixture of disodium 3,3'-dioxo-2,2'-bi-indolylidene-5,5'-disulfonate, and disodium 3,3'-dioxo-2,2'-bi-indolylidene-5,7'-disulfonate and subsidiary colouring matters together with sodium chloride and/or sodium sulfate as the principal uncoloured components. Indigotine may be converted to the corresponding aluminium lake in which case only the General Specifications for Aluminium Lakes of Colouring Matters shall apply.
Class	Indigoid
Code numbers	CI (1975) No.73015 CI (1975) Food Blue 1 CAS No. 860-22-0 (5,5' isomer) EEC No. E132
Chemical name	Disodium 3,3'-dioxo-2,2'-bi-indolylidene-5,5'-disulfonate
Chemical formula	$C_{16}H_8N_2Na_2O_8S_2$

Structural formula

$$\text{NaO}_3\text{S} - \underset{\underset{H}{N}}{\overset{O}{\bigcirc}} C = C \underset{\underset{H}{N}}{\overset{O}{\bigcirc}} - \text{SO}_3\text{Na}$$

Molecular weight	466.36
Assay	Content not less than 85% total colouring matters. Disodium 3,3'-dioxo-2,2'-bi-indolylidene-5,7'-disulfonate: not more than 18%.
DESCRIPTION	Blue powder or granules
FUNCTIONAL USE	Food colour

---

* Supersedes the earlier specification for Indigotine published in FAO Food and Nutrition Paper No 19 (1981).

CHARACTERISTICS

IDENTIFICATION TESTS

*A.   Solubility   Soluble in water
                          Sparingly soluble in ethanol

**B. Identification of colouring matters

PURITY TESTS

**Loss on drying at 135°   Not more than 15%

**Chloride and sulfate
  calculated as sodium salts   Not more than 15%

  Water insoluble matter   Not more than 0.2%
                                        (See description under TESTS)

**Subsidiary colouring matters   Excluding disodium 3,3'-dioxo-2,2'-
                                        bi-indolylidene-5,7'-disulfonate:
                                        not more than 1%.
                                        (See description under TESTS)

**Organic compounds other than colouring matters

    Isatin-5-sulfonic acid   ⎫
    5-sulfoanthranilic acid   ⎬   Total not more than 0.5%
    Anthranilic acid   ⎭   (See description under TESTS)

    Unsulfonated primary aromatic amines
                                        Not more than 0.01%
                                        calculated as aniline

**Ether extractable matter   Not more than 0.2%
                                        (See description under TESTS)

**Arsenic   Not more than 3 mg/kg

**Lead   Not more than 10 mg/kg

**Mercury   Not more than 1 mg/kg

*Heavy Metals   Not more than 40 mg/kg

TESTS

PURITY TESTS

**Water insoluble matter

Weigh accurately about 2 g sample instead of the 4.5-5 g stated in
the method.

---

*   See General Methods (Guide to Specifications, FAO Food and Nutrition
    Paper No 5, Revision 1, 1983).
** See Annex.

PURITY TESTS (cont'd)

*Subsidiary colouring matters

Use the following conditions:
  Developing solvent: No. 3
  Height of ascent of solvent front: approximately 17 cm

Note 1. The 5,7' isomer is separated as a wide blue zone just in front of the main blue band. Do not include this zone in the subsidiary colouring matter zones which are cut out and measured.

Note 2. The 15 ml sodium hydrogen carbonate solution used in the general procedure is replaced by 15 ml 0.05 N hydrochloric acid in order to avoid the decomposition which sulfonated indigo undergoes in alkaline solution.

*Organic compounds other than colouring matters

HPLC elution gradient:
  2 to 100% at 4.0% per minute (exponential)
  followed by elution at 100%.

*Ether extractable matter

Weigh accurately about 2 g sample instead of the 5 g stated in the method.

METHOD OF ASSAY

*Determination of Total Colouring Matters by Titration with Titanous Chloride

  Weight of sample: 1.0-1.1 g
  Buffer: 15 g sodium hydrogen tartrate
  Weight (D) of colouring matters equivalent to 1.00 ml of
    0.1 N $TiCl_3$: 0.02332 g

Isomer content by paper chromatography

Refer to the conditions for the determination of subsidiary colouring matters (above). Cut the isomer band from the chromatogram in the manner detailed for the subsidiary bands, extract into solvent and measure the absorbance at its λ max. Measure the absorbance of the corresponding blank at the same wavelength. As a standard use 0.1 ml of an 0.20% solution of the sample applied to the 18 cm x 0.7 cm rectangle.

Isomer expressed as a percentage of the sample = $\frac{A}{As} \times 20\% \times \frac{D}{100}$

where A and As are the net absorbances of the isomer and standard, respectively, and D is the total colouring matters content of the sample.

Isomer content by HPLC

The 5,7' isomer separates under the HPLC conditions detailed above for the separation of subsidiary colouring matters, and the amount present can be quantified using 5,7' isomer of known colouring matters content as the standard.

---

*   See Annex.

IRON OXIDES*

SYNONYM

DEFINITION Iron oxides are produced synthetically and consist essentially of anhydrous and/or hydrated iron oxides. The range of hues includes yellows, reds, browns and blacks. Food quality iron oxides are primarily distinguished from technical grades by the comparatively low levels of contamination by other metals. This is achieved by the selection and control of the source of the iron and/or by the extent of chemical purification during the manufacturing process.

    Class   Inorganic pigment

    Code numbers   CI (1975) No. 77192 (Iron oxide Yellow)
                                77491 (Iron oxide Red)
                                77499 (Iron oxide Black)
                  CI (1975) Pigment Yellow 42 and 43
                            Pigment Red 101 and 102
                            Pigment Black 11
                  CAS No. 20344-49-4 [FeO(OH)]
                            1309-33-7 [Fe(OH)$_3$]
                            1309-37-1 (Fe$_2$O$_3$)
                            1317-61-9 (Fe$_3$O$_4$)
                  EEC No. E172

    Chemical name   Hydrated ferric oxide (Iron oxide Yellow)
                            Anhydrous ferric oxide (Iron oxide Red)
                            Ferroso ferric oxide (Iron oxide Black)

    Chemical formula   FeO(OH) . x H$_2$O (Iron oxide Yellow)
                            Fe$_2$O$_3$ (Iron oxide Red)
                            Fe$_3$O$_4$ (Iron oxide Black)

    Molecular weight   88.85 [FeO(OH)]
                            159.70 (Fe$_2$O$_3$)
                            231.55 (Fe$_3$O$_4$)

    Assay   Not less than 60% of iron (Fe)

DESCRIPTION   Powder; yellow, red, brown or black in hue.

FUNCTIONAL USE   Food colour

CHARACTERISTICS

IDENTIFICATION TESTS

  A. Solubility   Insoluble in water
                        Insoluble in organic solvents
                        Solvent in concentrated mineral acids

---

*   Supersedes the earlier specifications for Iron Oxides published in the FAO Food and Nutrition Paper No 12 (1979).

**   See General Methods (Guide to Specifications, FAO Food and Nutrition Paper No 5, Revision 1, 1983).

PURITY TESTS

Water soluble matter	Not more than 1.0%	
Arsenic	Not more than 3 mg/kg	⎫
Lead	Not more than 10 mg/kg	⎪
Copper	Not more than 50 mg/kg	⎪
Zinc	Not more than 100 mg/kg	⎬ By total dissolution
Chromium	Not more than 100 mg/kg	⎪ (See description under TESTS)
Barium	Not more than 50 mg/kg	⎪
Nickel	Not more than 100 mg/kg	⎪
Cadmium	Not more than 10 mg/kg	⎪
Mercury	Not more than 1 mg/kg	⎭

TESTS

PURITY TESTS

Arsenic, lead, copper, zinc, chromium, barium, nickel, cadmium, mercury

Weigh 5 g of the sample and transfer to a beaker. Add 50 ml concentrated hydrochloric acid and heat on a hot plate until dissolved. Dilute with water to 100 ml in a volumetric flask. Determine the trace metals content by AAS.

METHOD OF ASSAY

Weigh accurately about 0.2 g of the sample, add 10 ml of 5 N hydrochloric acid and heat cautiously to boiling in a 200 ml conical flask until the sample has dissolved. Allow to cool, add 6 to 7 drops of 30% hydrogen peroxide solution and again heat cautiously to boiling until all the excess hydrogren peroxide has decomposed (about 2-3 minutes). Allow to cool, add 30 ml of water and about 2 g of potassium iodide and allow to stand for 5 minutes. Add 30 ml of water and titrate with 0.1 N sodium thiosulfate adding starch TS as the indicator towards the end of the titration. Each ml of 0.1 N sodium thiosulfate is equivalent to 5.585 mg of Fe (III).

LITHOL RUBINE BK*

SYNONYMS	D & C Red No.7 Brilliant Carmine 6B Rubinpigment Carmine 6B Litholrubintoner BKL Permanent Rubin L6B
DEFINITION	Lithol Rubine BK consists essentially of calcium salts of 3-hydroxy-4-(4-methyl-2-sulfophenylazo) naphthalene-2-carboxylic acid and subsidiary colouring matters together with calcium chloride and/or calcium sulfate as the principal uncoloured components.
Class	Monoazo
Code numbers	CI (1975) No.15850: 1 CI (1975) Pigment Red 57: 1 CAS No.5284-04-9 EEC No. E180
Chemical name	The calcium salt of 3-hydroxy-4-(4-methyl-2-sulfophenylazo) naphthalene-2-carboxylic acid
Chemical formula	$C_{18}H_{12}CaN_2O_6S$
Structural formula	(structure: CH_3-phenyl-$SO_3\frac{Ca}{2}$ — N=N — naphthalene with HO and $COO\frac{Ca}{2}$)
Molecular weight	424.44
Assay	Content not less than 90% total colouring matters
DESCRIPTION	Red powder
FUNCTIONAL USE	Food colour

CHARACTERISTICS

IDENTIFICATION TESTS

**A. Solubility	Slightly soluble in hot water (90°) Insoluble in cold water Insoluble in ethanol
B.	λmax 442 nm (in dimethylformamide)

---

* Supersedes the earlier specifications for Lithol Rubine BK published in FAO Food and Nutrition Paper No 25 (1982).

** See General Methods (Guide to Specifications, FAO Food and Nutrition Paper No 5, Revision 1, 1983).

PURITY TESTS

*Loss on drying at 135°	Not more than 10% (See description under TESTS)
Chloride and sulfate calculated as calcium salts	Not more than 10% (See description under TESTS)
Subsidiary colouring matters	Not more than 0.5% (See description under TESTS)

Organic compounds other than colouring matters

    2-amino-5-methylbenzene sulfonic acid, calcium salt
        Not more than 0.2%

    3-hydroxynaphthalene-2-carboxylic acid, calcium salt
        Not more than 0.4%
        (See description under TESTS)

    Unsulfonated primary aromatic amines
        Not more than 0.01% calculated as aniline
        (See description under TESTS)

Ether extractable matter	Not more than 0.2% (See description under TESTS)
*Arsenic	Not more than 3 mg/kg
*Lead	Not more than 10 mg/kg
**Heavy metals	Not more than 40 mg/kg

TESTS

PURITY TESTS

Chloride

Mix 1 g of the sample with 100 ml of water and let stand for 30 minutes, shaking occasionally. Filter and wash the residue with a small amount of water. Combine the washings with the filtrate. Acidify with 5 ml of 1.5 N nitric acid, and titrate with 0.1 N silver nitrate solution as directed in the Test for Chlorides.**

$$\text{\% of Chloride as calcium chloride} = \frac{\text{Amount of } 0.1\ N\ AgNO_3 \times 55.49}{1000} \times 100$$

Sulfate

Mix 1 g of the sample with 100 ml of water and heat on a water bath for 10 minutes. Cool, filter and wash the residue with a small amount of water. Combine the washings with the filtrate. Dilute to 150 ml with water, and acidify with hydrochloric acid, adding 1 ml in excess. Heat the solution to boiling, and add an excess of 0.25 N barium chloride, drop by drop, with stirring. Allow the mixture to stand on a hot plate

---

*   See Annex

**   See General Methods (Guide to Specifications, FAO Food and Nutrition Paper No 5, Revision 1, 1983).

PURITY TESTS (cont'd)

Sulfate (cont'd)

for 4 hours, or leave it overnight at room temperature and then bring it to about 80°, and allow the precipitate to settle. Filter off the precipitated barium sulfate, wash with hot water, and ignite at a dull red heat in a tared crucible until a constant weight is obtained. Carry out a blank determination and apply any necessary correction.

$$\% \text{ of Sulfate as calcium sulfate} = \frac{\text{Weight of BaSO}_4 \text{ (g)} \times 0.583}{1g} \times 100$$

Subsidiary colouring matters

Apparatus

- Spectrophotometer, suitable for use in the visible range
- Separatory funnels

Reagents

(a) Glacial acetic acid
(b) Diethyl ether
(c) Hydrochloric acid 8 N
(d) 2% Sodium hydroxide
(e) Standard solution of 1-(4-Tolylazo)-2-naphthol-3-carboxylic acid

Procedure

Boil 0.10 g of sample gently with 100 ml of glacial acetic acid and 75 ml of HCl 8 N until the colour has dissolved. Cool and transfer the solution to a 1000-ml separatory funnel, washing any residual solution into the funnel with small portions of acetic acid. Extract the acidic solution with 150 ml of ether, and separate the miscible solution formed by adding about 150 ml of water. Transfer the lower layer to a second funnel and extract with another 100 ml of ether.

Combine the ether extracts and wash with 100-ml portions of water until the washings are colourless, and twice more after the last colourless washing. Remove the subsidiary colour from the ether layer by extraction with 20-ml portions of 2% sodium hydroxide. Warm to expel the ether. Determine the colour concentration using the spectrophotometer.

Standard

3-hydroxy-4-(4-methylphenylazo) naphthalene-2-carboxylic acid has an absorbancy of 0.032/mg/l at 505 nm.

Organic compounds other than colouring matters

Use ammonium sulfate 10% in place of the ammonium sulfate 25% listed in the general method.

Add 100 ml eluant to 5 g cellulose, stir, allow to settle and decant. Place 0.100 g of the colour sample in a beaker and add 5 ml ethyl alcohol. Stir to ensure complete wetting of the sample. Transfer the 5 g cellulose to the beaker containing the sample. Add 10 g ammonium sulfate and stir thoroughly. Transfer the mixture to the column. Rinse the beaker with 25 ml eluant, adding the rinse to the column. Then follow the procedure detailed in the general column chromatography method.*

./.

---

* See Annex.

PURITY TESTS (cont'd)

Unsulfonated primary aromatic amines

   Mix 2 g of the sample well with 150 ml of toluene and boil gently for
   5 minutes. Filter after cooling and wash the residue with a small
   amount of toluene. Combine the washing with the filtrate, and extract
   with three 10-ml portions of 3 N hydrochloric acid and dilute the combined extract to 100 ml with water. Follow the procedure detailed in
   the general method.*

Ether extractable matter

   Dry 2 g of the sample in a desiccator (sulfuric acid) for 24 hours
   and extract, using Method II.*

METHOD OF ASSAY (Titration with Titanous Chloride)

   Place about 0.2 g of the sample, accurately weighed, in a 500-ml
   Erlenmeyer flask and add 5 ml of sulfuric acid. Mix well and add
   100 ml of ethanol. Shake well, heat on a water bath. Add a solution
   which dissolves 20 g of sodium hydrogen tartrate in 100 ml of boiling
   water and mix with 20 ml of 30% sodium hydroxide solution, shaking
   vigorously. Titrate with 0.1 N titanous chloride.

   Each ml of 0.1 N titanous chloride is equivalent to 10.61 mg of
   $C_{18}H_{12}N_2SO_6Ca$.

---

* See Annex.

PAPRIKA OLEORESIN*

SYNONYMS	Paprika extract
DEFINITION	Paprika oleoresin is obtained by solvent extraction of paprika, which consists of the ground fruit pods, with or without the seeds, of *Capsicum annuum L* and contains the major flavouring and colouring principles of this spice. Only the following solvents may be used: dichloromethane, trichloroethylene, acetone, propan-2-ol, methanol, ethanol, hexane. The solvent is subsequently removed. The major flavouring principle of paprika oleoresin is capsaicin. The major colouring principles of paprika oleoresin are capsanthin and capsorubin. A wide variety of other coloured compounds are known to be present.
Class	Carotenoid
Code numbers	CI (1975) - CI (1975) - CAS No. EEC No. E160c (capsanthin, capsorubin)
Chemical name	
Chemical formula	Capsaicin $C_{18}H_{27}NO_3$ Capsanthin $C_{40}H_{56}O_3$ Capsorubin $C_{40}H_{56}O_4$
Structural formula	Capsaicin:

$$CH_2NHCO(CH_2)_4CH=CH-CH(CH_3)_2$$

(phenyl ring with OH and OCH_3 substituents)

Capsanthin:

(structural diagram of capsanthin)

---

* Supersedes the earlier specifications for Paprika Oleoresins published in FAO Food and Nutrition Paper No 19 (1981).

Structural formula (cont'd)
Capsorubin:

Molecular weight	Capsanthin: 584.85
	Capsorubin: 600.85
Assay	Information required
DESCRIPTION	Dark red viscous liquid
FUNCTIONAL USE	Food colour, flavouring agent

CHARACTERISTICS

IDENTIFICATION TESTS

*A. Solubility   Practically insoluble in water
    Partially soluble with oily separation in ethanol
    Insoluble in glycerin

B. Spectrophotometry   In hexane the maximum absorption is about 470 nm.

C. Colour reaction   Passes test.
    (See description under TESTS)

PURITY TESTS

**Residual solvents

Dichloromethane / Trichloroethylene	Not more than 30 mg/kg - singly or in combination
Acetone	Not more than 30 mg/kg
Propan-2-ol	Not more than 50 mg/kg
Methanol	Not more than 50 mg/kg
Ethanol	Not more than 50 mg/kg
Hexane	Not more than 25 mg/kg
Capsaicin	Information required (See description under TESTS)

---

* See General Methods (Guide to Specifications, FAO Food and Nutrition Paper No 5, Revision 1, 1983).

** See Annex.

PURITY TESTS (cont'd)

Arsenic   Not more than 3 mg/kg

Lead   Not more than 10 mg/kg

Heavy metals   Not more than 40 mg/kg

TESTS

C. Colour reaction

To one drop of the sample add 2-3 drops of chloroform and one drop of sulfuric acid. A deep blue colour is produced.

Determination of Capsaicin

About 5 g are weighed exactly in a 300 ml ground joint flask. After addition of 100 ml methanol 70%, shake for 30 minutes. Let the solution settle for 5 minutes and filter. Cover the funnel to avoid evaporation. The first 25 ml of the filtrate are discarded and the rest of the filtrate mixed well. Afterwards, solutions are prepared in 100-ml flasks in the following manner:

	flask 1	flask 2	flask 3	flask 4
Solution	4.00 ml	4.00 ml	-	-
dist. H_2O	17.80 ml	16.80 ml	19.00 ml	18.00 ml
$1\underline{N}$ - HCl	1.00 ml	-	1.00 ml	-
$1\underline{N}$ - NaOH	-	2.00 ml	-	2.00 ml
determined value	E_1	E_2	E_3	E_4

The solutions are mixed well and the flasks filled to 100 ml with methanol p.a.

The extinction values E_1 - E_4 of the 4 solutions are measured at 248 nm and 296 nm (Deuterium lamp, quartz cuvettes).

Calculation:

(a) at 248 nm $\dfrac{[(E_2 - E_1) - (E_4 - E_3)] \cdot 2500}{314 \cdot \text{sample (in gram)}}$ = % Capsaicin

(b) at 296 nm $\dfrac{[(E_2 - E_1) - (E_4 - E_3)] \cdot 2500}{127 \cdot \text{sample (in gram)}}$ = % Capsaicin

2500 = dilution
314 and 127 = correction factors

The results of (a) and (b) must not differ more than 10%, otherwise the determination has to be repeated.

---

*   See Annex.
**   See General Methods (Guide to Specifications, FAO Food and Nutrition Paper No 5, Revision 1, 1983).

METHOD OF ASSAY

In the absence of an assay method a measurement of colour intensity, by the following method, may be used.

Apparatus

- Spectrophotometer, capable of accurately measuring absorbance at 460 nm.
- Absorption cells, 1 cm, square matched cells with stoppers.
- Volumetric flasks, 100 ml, with ground glass stoppers.
- Pipette, transfer-type, 10 ml.
- Glassine paper.
- Whatman No.40 filter paper or equivalent.

Reagents

- Acetone, technical grade.
- Cobaltous ammonium sulfate crystals.
- Potassium dichromate, reagent grade.

The cobaltous ammonium sulfate should be dried one week in a desiccator containing Drierite (anhydrous calcium sulfate).

No preliminary treatment is needed for the potassium dichromate.

Standard colour solution

0.3005 g/litre potassium dichromate plus 34.96 g/litre cobaltous ammonium sulfate crystals in 1.8$\underline{M}$ sulfuric acid solution.

Procedure

Weigh, on a 5 cm square of glassine paper, a sample of 50 to 80 mg to the nearest 0.1 mg.

Place the paper and the sample in the 100 ml volumetric flask and dilute to the mark with acetone.

Allow the extraction to proceed for at least 15 minutes with occasional shaking.

With a 10-ml pipette, transfer 10.0 ml of the extract into another 100-ml volumetric flask, and dilute to the mark with acetone.

Filter the diluted extract using Whatman No.40 filter paper or equivalent; discard the first 10 or 15 ml of filtrate.

Decant a portion of the filtrate into a cell and measure the absorbance at 460 nm using acetone as a blank.

Determine the absorbance (A_s) of the Standard colour solution at 460 nm.

Calculation

(a) Cell length and instrument correction factor = I_f

$I_f = 0.600/A_s$

(b) Extractable colour ASTA colour value =

$$\frac{\text{Absorbance of acetone extract at 460 nm} \times 164 \times I_f}{\text{sample weight in g}}$$

NOTES:
1. The absorbance of the standard colour solution (A_s) need only be determined once each day if the spectrophotomer is left on.

./.

NOTES (cont'd)

2. If cells having a path length other than 1 cm are used, the correction for the path length is contained within the factor I_f. Since the absolute response of instruments changes slightly every time the instrument is turned off and on, it is recommended that I_f be redetermined every day.

3. The recommended range of absorbance values is between $A = 0.30$ and $A = 0.70$. Extracts having $A > 0.70$ should be diluted with acetone to one half of the original concentration. Extracts having $A < 0.30$ should be discarded and the extraction performed again using a larger sample.

PATENT BLUE V*

SYNONYMS	CI Food Blue 5 Patent Blue 5
DEFINITION	Patent Blue V consists essentially of the calcium or sodium compound of [4-[α-(4-diethylaminophenyl)-5-hydroxy-2,4 disulfonphenyl-methlidene]2,5-cyclohexadien-1-ylidene] diethyl-ammonium hydroxide inner salt and subsidiary colouring matters together with sodium chloride and/or sodium sulfate and/or calcium chloride and/or calcium sulfate as the principal uncoloured components. Patent Blue V may be converted to the corresponding aluminium lake in which case only the General Specifications for Aluminium Lakes of Colouring Matters shall apply.
Class	Triarylmethane
Code numbers	CI (1975) No.42051 CI (1975) Food Blue 5 CAS No. 3536-49-0 EEC No. E131
Chemical name	The calcium or sodium compound of [4-[α-(4-diethyl-aminophyenly)-5-hydroxy-2,4 disulfophenyl-methylidene] 2,5-cyclohexadien-1-ylidene] diethyl-ammonium hydroxide inner salt
Chemical formula	Calcium compound: $\{C_{27}H_{31}N_2O_7S_2\}_2 Ca$ Sodium compound: $C_{27}H_{31}N_2O_7S_2Na$
Structural formula	Calcium compound

[Structural formula image showing the calcium compound with HO, $\frac{Ca}{2}O_3S$, SO_3^-, $N(C_2H_5)_2$, and $=N^+(C_2H_5)_2$ groups]

Molecular weight	Calcium compound: 579.73 Sodium compound: 582.66
Assay	Content not less than 85% total colouring matters
DESCRIPTION	Blue powder or granules
FUNCTIONAL USE	Food Colour

---

* Supersedes the earlier specifications for Patent Blue V published in FAO Food and Nutrition Paper No 25 (1982).

CHARACTERISTICS

IDENTIFICATION TESTS

*A. Solubility   Soluble in water
                        Slightly soluble in ethanol

**B. Identification of colouring matters

PURITY TESTS

**Loss on drying at 135°	Not more than 15%
**Chloride and sulfate calculated as sodium salts	Not more than 15%
**Water insoluble matter	Not more than 0.2%
Subsidiary colouring matters	Not more than 2% (See description under TESTS)

Organic compounds other than colouring matters

3-hydroxy benzaldehyde 3-hydroxy benzoic acid 3-hydroxy-4-sulfobenzoic acid N,N-diethylamino benzene sulfonic acids	Total not more than 0.5% (See description under TESTS)
Leuco base	Not more than 4% (See description under TESTS)
Unsulfonated primary aromatic amines	Not more than 0.01% calculated as aniline
**Ether extractable matter	Not more than 0.2%
**Arsenic	Not more than 3 mg/kg
**Lead	Not more than 10 mg/kg
**Mercury	Not more than 1 mg/kg
**Chromium	Not more than 50 mg/kg
*Heavy metals	Not more than 40 mg/kg

TESTS

PURITY TESTS

Subsidiary colouring matters

Use the following conditions:
    Developing solvent: No.2
    Height of ascent of solvent front: approximately 17 cm

---

* See General Methods (Guide to Specifications, FAO Food and Nutrition Paper No 5, Revision 1, 1983).

** See Annex.

PURITY TESTS (cont'd)

Organic compounds other than colouring matters

Use the following conditions:

Instrument:	High Performance Liquid Chromatograph fitted with a gradient elution accessory
Detector:	A UV HPLC detector recording absorbances at 254 nm.
Column:	250 x 4 mm (Kartusche). Li Chrosorb RP 18, 7 μm
Solvent system:	(a) Acetate buffer pH 4.6:water (1:10)* (b) Acetonitrile

Gradient:

Minutes	%(a)	%(b)	Flow rate (ml/min.)
0	85	15	1
12	85	15	1
25	20	80	2
28	20	80	2
40	85	15	1

METHOD OF ASSAY

Determination of Total Colouring Matters by Titration with Titanous Chloride

Use the following:

Weight of sample: 1.3-1.4 g
Buffer: 15 g sodium hydrogen tartrate
Weight (D) of colouring matters equivalent to 1.00 ml of 0.1 N TiCl$_3$: 0.02898 g of the calcium salt, 0.02913 g of sodium salt.

---

* Acetate buffer is Sodium hydroxide 1 $\underline{M}$: Acetic acid 1 $\underline{M}$: water (5:10:35).

PONCEAU 4R*

SYNONYMS	CI Food Red 7 Cochineal Red A New Coccine
DEFINITION	Ponceau 4R consists essentially of trisodium-2-hydroxy-1-(4-sulfonato-1-naphthylazo) naphthalene-6,8-disulfonate, and subsidiary colouring matters together with sodium chloride and/or sodium sulfate as the principal uncoloured components. Ponceau 4R may be converted to the corresponding aluminium lake in which case only the General Specifications for Aluminium Lakes of Colouring Matters shall apply.
Class	Monoazo
Code numbers	CI (1975) No. 16255 CI (1975) Food Red 7 CAS No. 2611-82-7 EEC No. E124
Chemical name	Trisodium-2-hydroxy-1-(4-sulfonato-1-naphthylazo) naphthalene-6,8-disulfonate.
Chemical formula	$C_{20}H_{11}N_2Na_3O_{10}S_3$
Structural formula	

$$NaO_3S-\text{[naphthyl]}-N=N-\text{[naphthyl(OH)(SO_3Na)(SO_3Na)]}$$

Molecular weight	604.48
Assay	Content not less than 80% total colouring matters
DESCRIPTION	Reddish powder or granules
FUNCTIONAL USE	Food colour

CHARACTERISTICS

IDENTIFICATION TESTS

**A. Solubility	Soluble in water Sparingly soluble in ethanol

---

* Supersedes the earlier specifications for Ponceau 4R published in FAO Food and Nutrition Paper No 19 (1981).

** See General Methods (Guide to Specifications, FAO Food and Nutrition Paper No 5, Revision 1, 1983).

IDENTIFICATION TESTS (cont'd)

*B. Identification of colouring matters

PURITY TESTS

*Loss on drying at 135°	Not more than 20%
*Chloride and sulfate calculated as sodium salts	Not more than 20%
*Water insoluble matter	Not more than 0.2%
Subsidiary colouring matters	Not more than 1% (See description under TESTS)

Organic compounds other than colouring matters

> 4-amino naphthalene-1-sulfonic acid
> 7-hydroxy naphthalene-1,3-disulfonic acid
> 3-hydroxy naphthalene-2,7-disulfonic acid
> 6-hydroxy naphthalene-2-sulfonic acid
> 7-hydroxy naphthalene-1,3,6-trisulfonic acid

Total not more than 0.5%
(See description under TESTS)

*Unsulfonated primary aromatic amines	Not more than 0.01% calculated as aniline
*Ether extractable matter	Not more than 0.2%
*Arsenic	Not more than 3 mg/kg
*Lead	Not more than 10 mg/kg
**Heavy metals	Not more than 40 mg/kg

TESTS

PURITY TESTS

*Subsidiary colouring matters

Use the following conditions:
Developing solvent: No. 3
Height of ascent of solvent front: 17 cm, then one hour further development

*Organic compounds other than colouring matters

Use HPLC under the following conditions:
HPLC elution gradient: 2 to 100% at 4% per minute (linear)

METHOD OF ASSAY

*Determination of Total Colouring Matters by Titration with Titanous Chloride

Use the following:
Weight of sample: 0.7-0.8 g
Buffer: 10 g sodium citrate
Weight (D) of colouring matters equivalent to 1.00 ml of 0.1 N $TiCl_3$: 0.01511 g

---

* See Annex.
** See General Methods (Guide to Specifications, FAO Food and Nutrition Paper No 5, Revision 1, 1983).

QUINOLINE YELLOW*

SYNONYMS	CI Food Yellow 13
DEFINITION	Quinoline Yellow is prepared by sulfonating 2-(2-quinolyl) indan-1,3-dione or a mixture continaing about two-thirds 2-(2-quinolyl) indan-1,3-dione and one-third 2-[2-(6-methyl=quinolyl)] indan-1,3-dione. Quinoline Yellow consists essentially of sodium salts of a mixture of disulfonates (principally), monosulfonates and trisulfonates of the above compounds and subsidiary colouring matters together with sodium chloride and/or sodium sulfate as the principal uncoloured components.
	Quinoline Yellow may be converted to the corresponding aluminium lake, in which case only the General Specifications for Aluminium Lakes of Colouring Matters shall apply.
Class	Quinophthalone
Code numbers	CI (1975) No. 47005 CI (1975) Food Yellow 13 CAS No. 8004-72-0 (Unmethylated disulfonic acids) EEC No. E104
Chemical name	The disodium salts of the disulfonates of 2-(2-quinolyl) indan-1,3-dione (principal components).
Chemical formula	$C_{18}H_9N\ Na_2O_8S_2$ (principal components).
Structural formula	(principal components)

$(NaO_3S)_2$ — [quinoline-indandione structure]

Molecular weight	477.38 (principal components)
Assay	Content not less than 70% total colouring matters.
	Quinoline Yellow prepared from 2-(2-quinolyl) indan-1,3-dione (only) shall have the following composition:
	Of the total colouring matters present: - Not less than 80% shall be disodium 2-(2-quinolyl) indan-1,3-dione-disulfonates; - Not more than 15% shall be sodium 2-(2-quinolyl) indan-1,3-dione-monosulfonates; - Not more than 7% shall be trisodium 2-(2-quinolyl) indan-1,3-dione-trisulfonate.

---

* Supersedes the earlier specification for Quinoline Yellow published in FAO Food and Nutrition Paper No 25 (1982).

DESCRIPTION   Yellow powder or granules

FUNCTIONAL USE   Food colour

CHARACTERISTICS

  IDENTIFICATION TESTS

*A. Solubility   Soluble in water
                             Sparingly soluble in ethanol

**B. Identification of colouring matters

  PURITY TESTS

**Loss on drying at 135°   Not more than 30%

**Chloride and sulfate
   calculated as sodium salts   Not more than 30%

**Water insoluble matter   Not more than 0.2%

Subsidiary colouring matters   2-(2-quinolyl) indan-1,3-dione and
                                 2-[2-(6-methylquinolyl)] indan-1,3-dione:
                                 Not more than 4 mg/kg
                                 (See description under TESTS)

Organic compounds other than colouring matters

    2-methylquinoline
    2-methylquinoline-sulfonic acid
    Phthalic acid   Total not more than 0.5%
*** 2,6-dimethyl quinoline   (See description under TESTS)
*** 2,6-dimethyl quinoline sulfonic acid

  ** Unsulfonated primary aromatic amines:
                                 Not more than 0.01%
                                 calculated as aniline

** Ether extractable matter   Not more than 0.2%

** Arsenic   Not more than 3 mg/kg

** Lead   Not more than 10 mg/kg

** Mercury   Not more than 1 mg/kg

** Zinc   Not more than 50 mg/kg

* Heavy metals   Not more than 40 mg/kg

---

*   See General Methods (Guide to Specifications, FAO Food and Nutrition Paper No 5, Revision 1, 1983).

**   See Annex.

***   Only in Quinoline Yellow prepared from the mixture of 2-(2-quinolyl)=indan-1,3-dione and 2-[2-(6-methylquinolyl)] indan-1,3-dione.

TESTS

PURITY TESTS

Subsidiary colouring matters

Limit test for 2-(2-quinolyl) indan-1,3-dione and 2-[2-(6-methyl=quinolyl)] indan-1,3-dione:

Use the apparatus and ether quality described in the Method for the Determination of Ether Extractable Matter* and carry out an extraction following the details given thereunder. Wash the ether extract with two 25-ml portions of water. Evaporate the ether extract to about 5 ml and then transfer it to an oven at 105°C to remove the remaining ether. Dissolve the residue in chloroform, and dilute the solution to exactly 20 ml. Determine the absorbance at the wavelength of maximum absorption (approximately 420 nm) using chloroform as the reference solution. The absorbance corresponding to the limit figure of 4 mg/kg 2-(2-quinolyl) indan-1,3-dione is 0.500. Any 2-[2-(6-methylquinolyl)] indan-1,3-dione is assessed as 2-(2-quinolyl) indan-1,3-dione.

Organic compounds other than colouring matters in Quinoline Yellow prepared from 2-(2-quinolyl) indan-1,3-dione (only)

Use the following conditions:

Instrument:	High performance liquid chromatograph fitted with a gradient elution accessory.
Detector:	A UV HPLC detector recording absorbances at 254 nm
Column:	250 x 4 mm, Nucleosil C_{18}, 7 μm
Solvent system:	A)**Acetate Buffer pH 4·6: water (1:10) B) (A):methanol (20:80)
Sample concentration:	1% weight/volume in Solvent A

Gradient:

Minutes	%A	%B
0	100	0
15	65	35
20	50	50
25	0	100
36	0	100
42	100	0

Flow rate: 1 ml/minute

METHOD OF ASSAY

Determination of Total Colouring Matters by Spectrophotometry

Solvent: pH 7 phosphate buffer
Dilution of solution A: 10 ml → 250 ml
Absorptivity (a): 86.5
Approximate wavelength of maximum absorption: 415 nm

Determination of the percentages of di-, mono-, and trisulfonates in Quinoline Yellow prepared from 2-(2-quinolyl) indan-1,3-dione (only)

Use the HPLC conditions prescribed in the Determination of Organic Compounds other than Colouring Matters with a sample solution of concentration 0.05% in HPLC Solvent A in place of the sample solution of concentration 1%. Express the results as percentages of the total colouring matters present.

---

* See Annex.
** Acetate buffer is sodium hydroxide 1M:Acetic acid 1M:water (5:10:35).

RED 2G*

SYNONYMS
: CI Food Red 10
Azogeranine

DEFINITION
: Red 2G consists essentially of disodium 8-acetamido-1-hydroxy-2-phenylazonaphthalene-3,6-disulfonate and subsidiary colouring matters together with sodium chloride and/or sodium sulfate as the principal uncoloured components.

  Red 2G may be converted to the corresponding aluminium lake in which case only the General Specifications for Aluminium Lakes of Colouring Matters shall apply.

Class
: Monoazo

Code numbers
: CI (1975) No. 18050
CI (1975) Food Red 10
CAS No. 3734-67-6
EEC No. 128

Chemical name
: Disodium 8-acetamido-1-hydroxy-2-phenylazonaphtalene-3,6-disulfonate

Chemical formula
: $C_{18}H_{13}N_3Na_2O_8S_2$

Structural formula
:

Molecular weight
: 509.43

Assay
: Content not less than 80% total colouring matters.

DESCRIPTION
: Red powder or granules

FUNCTIONAL USE
: Food colour

CHARACTERISTICS

IDENTIFICATION TESTS

**A. Solubility
: Soluble in water
Sparingly soluble in ethanol

***B. Identification of colouring matters

---

* Supersedes the earlier specifications for Red 2G published in FAO Food and Nutrition Paper No 19 (1981).

** See General Methods (Guide to Specifications, FAO Food and Nutrition Paper No 5, Revision 1, 1983).

*** See Annex.

PURITY TESTS

*Loss on drying at 135°	Not more than 20%
*Chloride and sulfate calculated as sodium salts	Not more than 20%
*Water insoluble matter	Not more than 0.2%
Subsidiary colouring matters	Not more than 2% (See description under TESTS)

Organic compounds other than colouring matters

    5-acetamido-4-hydroxynaphthalene-2,7-disulfonic acid
    5-amino-4-hydroxynaphthalene-2,7-disulfonic acid

Total not more than 0.3% (See description under TESTS)

  *Unsulfonated primary aromatic amines: Not more than 0.01% calculated as aniline

*Ether extractable matter	Not more than 0.2%
*Arsenic	Not more than 3 mg/kg
*Lead	Not more than 10 mg/kg
**Heavy metals	Not more than 40 mg/kg

TESTS

PURITY TESTS

*Subsidiary colouring matters

    Use the following conditions:
        Developing solvent: No. 4
        Height of ascent of solvent front: approximately 17 cm

*Organic compounds other than colouring matters

    Use HPLC under the following conditions:
        HPLC elution gradient: 1 to 100% at 2.5% per minute (linear)

METHOD OF ASSAY

*Determination of Total Colouring Matters by Titration with Titanous Chloride

    Weight of sample: 0.6-0.7 g
    Buffer: 15 g sodium hydrogen tartrate
    Weight (D) of colouring matters equivalent to 1.00 ml of 0.1 N $TiCl_3$: 0.01274 g

---

*   See Annex.
**   See General Methods (Guide to Specifications, FAO Food and Nutrition Paper No 5, Revision 1, 1983).

RIBOFLAVIN*

SYNONYMS	Vitamin B_2 Lactoflavine
DEFINITION	
Class	Iso-alloxazine
Code numbers	CAS No. 83-88-5 EEC No. E101
Chemical name	7,8-dimethyl-10-(1'-D-ribityl)isoalloxazine
Chemical formula	$C_{17}H_{20}N_4O_6$
Structural formula	
Molecular weight	376.37
Assay	Content not less than 98% of total colouring matters
DESCRIPTION	Yellow to orange-yellow powder, with slight odour
FUNCTIONAL USE	Food colour
CHARACTERISTICS	

IDENTIFICATION TESTS

**A. Solubility Very slightly soluble in water.
Practically insoluble in alcohol, chloroform, acetone and ether.
Very soluble in dilute alkali solutions.

B. Using the aqueous solution from the Assay, determine the absorbance at 267 nm, 375 nm and 444 nm.

The ratio A_{375}/A_{267} is between 0.31 and 0.33.

The ratio A_{444}/A_{267} is between 0.36 and 0.39.

C. Specific rotation, $[\alpha]_D^{20°}$ between $-115°$ and $-140°$.
(See description under TESTS)

D. Passes test
(See description under TESTS)

---

* Supersedes the earlier specifications for Riboflavin published in FAO Nutrition Meeting Report Series (1959) Vol. II.

** See General Methods (Guide to Specifications, FAO Food and Nutrition Paper No 5, Revision 1, 1983).

PURITY TESTS

*Loss on drying at 105°	Not more than 1.5% after 4 hours.
**Sulfated ash	Not more than 0.1% (See description under TESTS)
Subsidiary colouring matters	
Lumiflavin	Passes test (See description under TESTS)
*Arsenic	Not more than 3 mg/kg
*Lead	Not more than 10 mg/kg
**Heavy metals	Not more than 40 mg/kg

TESTS

IDENTIFICATION TESTS

C. Specific rotation $[\alpha]_D^{20°}$

Dry the sample at 105° for 4 hours. Dissolve 50.0 mg in 0.05 N sodium hydroxide free from carbonate and dilute to 10.0 ml with the same solvent. Measure the optical rotation within 30 minutes of dissolution.

D. Dissolve about 1 mg in 100 ml of water. The solution has a pale greenish-yellow colour by transmitted light, and by reflected light has an intense yellowish-green fluorescence which disappears on the addition of mineral acids and alkalis.

PURITY TESTS

Sulfated Ash

Proceed as directed under the test of Ash** using 2.0 g of the sample.

Lumiflavin

Shake 25 mg with 10 ml of ethanol-free chloroform R for 5 minutes and filter. The filtrate is not more intensely coloured than reference solution Y_6.

Standard Solution Y is prepared as follows:
Dissolve 46 g of $FeCl_3.6H_2O$ in about 900 ml of a mixture of 25 ml of concentrated hydrochloric acid and 975 ml of water and dilute to 1000.0 ml with the same mixture. Titrate the solution and adjust it to contain 45.0 mg of $FeCl_3.6H_2O$ per ml by adding the dilute hydrochloric acid mixture.

Titration:
Place in a 200-ml conical flask fitted with a ground-glass stopper, 10.0 ml of the solution, 15 ml of water, 5 ml of hydrochloric acid R and 4 g of potassium iodide R, close the flask, allow to stand in the dark for 15 minutes and add 100 ml of water. Titrate the liberated iodine with 0.1 N sodium thiosulphate, using 10 drops of

./.

---

* See Annex.

** See General Methods (Guide to Specifications, FAO Food and Nutrition Paper No 5, Revision 1, 1983).

Titration (cont'd)

starch solution R as indicator, added towards the end of the titration. 1 ml of 0.1 N sodium thiosulfate is equivalent to 27.03 mg of $FeCl_3.6H_2O$.

Solution Y_6 is prepared by dilution of 0.1 ml of Standard Solution Y with 1% hydrochloric acid.

METHOD OF ASSAY

Carry out the assay in subdued light. In a brown glass 500-ml volumetric flask, suspend 65.0 mg of the sample in 5 ml of water, ensuring that it is completely wetted, and dissolve in 5 ml of 2 N sodium hydroxide solution. As soon as dissolution is complete, add 100 ml of water and 2.5 ml of glacial acetic acid and dilute to 500.0 ml with water. Place 20.0 ml of this solution in a brown glass 200-ml volumetric flask, add 3.5 ml of a 1.4% m/V solution of sodium acetate and dilute to 200.0 ml with water. Measure the absorbance at the maximum at 444 nm. Calculate the content of $C_{17}H_{20}N_4O_6$ taking the specific absorbance to be 328.

RIBOFLAVIN 5'-PHOSPHATE SODIUM*

SYNONYMS
: Riboflavin 5'-phosphate ester monosodium salt; Vitamin B_2 phosphate ester monosodium salt.

DEFINITION

Class
: Iso-alloxazine

Code numbers
: CAS No. 130-40-5
  EEC No. 101a

Chemical name
: Monosodium salt of the 5'-monophosphate ester of riboflavin.

Chemical formula
: $C_{17}H_{20}N_4O_9P$ $Na \cdot 2H_2O$

Structural formula

Molecular weight
: 514.37

Assay
: Content not less than 70% total colouring matters.

DESCRIPTION
: Yellow to orange crystalline hygroscopic powder, with a slight odour and a bitter taste.

FUNCTIONAL USE
: Food colour

CHARACTERISTICS

IDENTIFICATION TESTS

**A. Solubility
: Soluble in water
  Insoluble in ethanol

**B. Specific rotation $[\alpha]_D^{20°}$
: Not less than +38° and not more than +42°
  (See description under TESTS)

**C. Sodium
: Passes test
  (See description under TESTS)

D.
: The ratio of the absorbances in aqueous solution at the maxima of 375 nm and 267 nm is between 0.30 and 0.34 and that of the absorbances at the maxima of 444 nm and 267 nm is between 0.35 and 0.40.

---

* Supersedes the earlier specification for Riboflavin 5'-Phosphate Sodium published in FAO Food and Nutrition Paper No 19 (1981).

** See General Methods (Guide to Specifications, FAO Food and Nutrition Paper No 5, Revision 1, 1983).

CHARACTERISTICS (cont'd)

PURITY TESTS

Loss on drying at 100°	Not more than 8.0% (See description under TESTS)
Sulfated ash	Not more than 25% (See description under TESTS)
Inorganic phosphate	Not more than 1% calculated as PO_4 on a dried basis (See description under TESTS)
Subsidiary colouring matters	
Riboflavin (free)	Not more than 6% (See description under TESTS)
Riboflavine diphosphate	Not more than 6% (See description under TESTS)
Lumiflavin	Not more than 250 mg/kg (See description under TESTS)
*Arsenic	Not more than 3 mg/kg
*Lead	Not more than 10 mg/kg
**Heavy metals	Not more than 40 mg/kg

TESTS

Conduct all tests and assays in light-resistant containers and protect the sample from direct sunlight at all stages.

IDENTIFICATION TESTS

B. Specific rotation $[\alpha]_D^{20°}$

Dry the sample at 100° for 5 hours in a vacuum over phosphorus pentoxide. Prepare a 1.5% weight/volume solution in 20% weight/volume hydrochloric acid and determine the specific rotation.**

C. Sodium

Use the sulfated ash and test for sodium as directed in General Methods.**

PURITY TESTS

Loss on drying at 100°

Dry the sample for 5 hours in a vacuum over phosphorus pentoxide.

Sulfated ash

Proceed as directed under the test of Ash* using 0.5 g of the sample.

---

*   See Annex.

** See General Methods (Guide to Specifications, FAO Food and Nutrition Paper No 5, Revision 1, 1983).

PURITY TESTS (cont'd)

Inorganic phosphate

Standard preparation
Transfer 220.0 mg of monobasic potassium phosphate KH_2PO_4, to a 1000-ml volumetric flask, dissolve in and dilute to volume with water and mix. Transfer 20.0 ml of this solution to a 100-ml volumetric flask, dilute to volume with water and mix.

Test preparation
Transfer 300.0 mg of the sample to a 100-ml volumetric flask, dissolve in and dilute to volume with water, and mix.

Acid molybdate solution
Dilute 25 ml of ammonium molybdate solution [7 g of $(NH_4)_6Mo_7O_{24} \cdot 4H_2O$] in sufficient water to make 100 ml to 200 ml with water, and then add slowly 25 ml of 7.5 $\underline{N}$ sulfuric acid.

Ferrous sulfate solution
Just before use, prepare a 10% aqueous ferrous sulfate solution containing 2 ml of 7.5 $\underline{N}$ sulfuric acid per 100 ml of final solution.

Procedure
Transfer 10.0 ml each of the Standard preparation and of the Test preparation into separate 50-ml Erlenmeyer flasks, add 10.0 ml of Acid molybdate solution and 5.0 ml of Ferrous sulfate solution to each flask, and mix. Determine the absorbance of each solution in a 1 cm cell at 700 nm with a suitable spectrophotometer, using as the blank a mixture of 10.0 ml of water, 10.0 ml of Acid molybdate solution, and 5.0 ml of Ferrous sulfate solution. The absorbance of the solution from the Test preparation is not greater than that of the Standard preparation.

Free riboflavin and Riboflavin diphosphate

Standard preparation
Transfer 35.0 mg of U.S.P. Riboflavin reference standard into a 250-ml Erlenmeyer flask, add 20 ml of pyridine and 75 ml of water, and dissolve the riboflavin by frequent shaking. Transfer the solution to a 1000-ml volumetric flask, dilute to volume with water, and mix. Transfer 20.0 ml of this solution to a second 1000-ml volumetric flask, adjust the pH to 6.0 by the addition of 8 ml of 0.1 $\underline{N}$ sulfuric acid, dilute to volume with water, and mix. Finally, transfer 25.0 ml of the last solution into a 100-ml volumetric flask, dilute to volume with dioxane-water mixture (1:3), and mix. This solution contains 0.175 μg of riboflavin per ml.

pH Buffer solution
Dissolve 15.6 g of monobasic sodium phosphate ($NaH_2PO_4 \cdot 2H_2O$) in about 100 ml of water, add 59.3 ml of 1 $\underline{N}$ sodium hydroxide TS, and dilute to 2000 ml with water. Check the pH with a pH meter, and adjust to 7.0 if necessary.

Test preparation
Dissolve 100.0 mg of the sample in 10.0 ml of pH7 Buffer solution. Prepare a strip of Whatman chromatography paper, Type 3 mm, medium flow rate, or other equivalent paper suitable for electrophoresis, and saturate the paper with pH7 Buffer solution. Using a micropipette, apply 0.01 ml of the sample solution along a narrow line of the cathode side of the paper strip contained in a suitable paper electrophoresis chamber. Apply a potential of approximately 250 v, allow electrophoresis to continue for 6 hours, and then remove the paper from the chamber

./.

Test preparation (cont'd)

Detect any free riboflavin and/or riboflavin diphosphate by observing the strip in daylight or under ultraviolet light. Free riboflavin, if present, will appear as a band nearest to the starting line, and riboflavin diphosphate will appear farthest from the starting line.

CAUTION: The riboflavin will be destroyed if exposed to the ultraviolet light for more than a few seconds.

Cut off the respective bands, place them in separate 250-ml Erlenmeyer flasks containing 35.0 ml of dioxane-water mixture (1:3), and allow to stand until the spots are completely eluted from the strips.

Procedure

Using a suitable fluorometer, determine the intensity of the fluorescence of each sample solution and of the Standard preparation at about 460 nm. The fluorescence of the sample solution containing the eluted riboflavin band and riboflavin diphosphate band, respectively, is not greater than that produced by the Standard preparation.

Lumiflavin

Prepare the standard for this limit test by diluting 3 ml of 0.1 $\underline{N}$ potassium dichromate with water to 1000 ml.

Pour some chloroform through an alumina column to remove any ethanol. To 10 ml of this chloroform add 35 mg of the sample, shake for 5 minutes and filter. The colour of the filtrate should be no more intense than that of 10 ml of the standard when viewed in identical containers.

METHOD OF ASSAY

Sample Preparation

Transfer about 50 mg of the sample, accurately weighed, into a 250-ml Erlenmeyer flask, add 20 ml of pyridine and 75 ml of water, and dissolve the sample by frequent shaking. Transfer the solution to a 1000-ml volumetric flask, dilute to volume with water, and mix. Transfer 10.0 ml of this solution into a second 1000-ml volumetric flask, add sufficient 0.1 $\underline{N}$ sulfuric acid (about 4 ml) so that the final pH of the solution is between 5.9 and 6.1, dilute to volume with water, and mix.

Standard Preparation

Transfer about 35 mg of U.S.P. Riboflavin reference standard, accurately weighed, into a 250-ml Erlenmeyer flask, add 20 ml of pyridine and 75 ml of water, and dissolve the riboflavin by frequent shaking. Transfer the solution to a 1000-ml volumetric flask, dilute to volume with water, and mix. Transfer 10.0 ml of this solution to a second 1000-ml volumetric flask, add sufficient 0.1 $\underline{N}$ sulfuric acid (about 4 ml) so that the final pH of the solution is between 5.9 and 6.1; dilute to volume with water, and mix.

Procedure

Using a suitable fluorometer, determine the intensity of the fluorescence of each solution at about 460 nm, recording the fluorescence of the sample preparation as I_u and that of the Standard preparation as I_s.

Calculation

Calculate the quantity, in mg, of $C_{17}H_{20}N_4O_6$ in the sample taken by the formula:

$$100C \times I_u / I_s$$

in which C is the exact concentration, in μg per ml, of the Standard preparation, corrected for loss on drying (105°, 2 hours).

SUNSET YELLOW FCF*

SYNONYMS	CI Food Yellow 3 FD&C Yellow No. 6 Crelborange S
DEFINITION	Sunset Yellow FCF consists essentially of disodium 2-hydroxy-1-(4-sulfonatophenylazo) naphthalene-6-sulfonate and subsidiary colouring matters together with sodium chloride and/or sodium sulfate as the principal uncoloured components. Sunset Yellow FCF may be converted to the corresponding aluminium lake in which case only the General Specifications for Aluminium Lakes of Colouring Matters shall apply.
Class	Monoazo
Code numbers	CI (1975) No. 15985 CI (1975) Food Yellow 3 CAS No. 2783-94-0 EEC No. E110
Chemical name	Disodium 2-hydroxy-1-(4-sulfonatophenylazo) napthalene-6-sulfonate
Chemical formula	$C_{16}H_{10}N_2Na_2O_7S_2$
Structural formula	
Molecular weight	452.37
Assay	Content not less than 85% total colouring matters
DESCRIPTION	Orange-red powder or granules
FUNCTIONAL USE	Food colour
CHARACTERISTICS	

IDENTIFICATION TESTS

**A. Solubility Soluble in water
              Sparingly soluble in ethanol

***B. Identification of colouring matters

---

* Supersedes the earlier specifications for Sunset Yellow FCF published in FAO Nutrition Meetings Report Series No 25 (1982).

** See General Methods (Guide to Specifications, FAO Food and Nutrition Paper No 5, Revision 1, 1983).

*** See Annex.

PURITY TESTS

*Loss on drying at 135°	Not more than 15%
*Chloride and sulfate calculated as sodium salts	Not more than 15%
*Water insoluble matter	Not more than 0.2%
Subsidiary colouring matters	Not more than 5% (See description under TESTS)
	Not more than 2% shall be colours other than trisodium 2-hydroxy-1-(4-sulfonatophenylazo) naphthalene-3,6-disulfonate

Organic compounds other than colouring matters

    4-aminobenzene-1-sulfonic acid
    3-hydroxynaphthalene-2,7-disulfonic acid
    6-hydroxynaphthalene-2-sulfonic acid   Total not more than 0.5%
    7-hydroxynaphthalene-1,3-disulfonic acid
    4,4'-diazoaminodi (benzenesulfonic acid)   (See description under TESTS)
    6,6'-oxydi(naphthalene-2-sulfonic acid)

    Unsulfonated primary aromatic amines:

	Not more than 0.01% calculated as aniline
*Ether extractable matter	Not more than 0.2%
*Arsenic	Not more than 3 mg/kg
*Lead	Not more than 10 mg/kg
**Heavy metals	Not more than 40 mg/kg

TESTS

  PURITY TESTS

  *Subsidiary colouring matters

    Use the following conditions:
      Developing solvent: No. 4
      Height of ascent of solvent front: approximately 17 cm

  *Organic compounds other than colouring matters

    Use HPLC under the following conditions:
      HPLC elution gradient: 2 to 100% at 4% per minute (linear) followed by elution at 100%.

METHOD OF ASSAY

  *Determination of Total Colouring Matters by Titration with Titanous Chloride

    Use the following:
      Weight of sample: 0.5-0.6 g
      Buffer: 10 g sodium citrate
      Weight (D) of colouring matters equivalent to 1.00 ml of 0.1 N $TiCl_3$: 0.01131 g.

---

*   See Annex.

** See General Methods (Guide to Specifications, FAO Food and Nutrition Paper No 5, Revision 1, 1983).

TARTRAZINE*

SYNONYMS	CI Food Yellow 4 FD&C Yellow No 5
DEFINITION	Tartrazine consists essentially of trisodium 5-hydroxy-1-(4-sulfonatophenyl)-4-(4-sulfonatophenylazo) pyrazole-3-carboxylate and subsidiary colouring matters together with sodium chloride and/or sodium sulfate as the principal uncoloured components. Tartrazine may be converted to the corresponding aluminium lake in which case only the General Specifications for Aluminium Lakes of Colouring Matters shall apply.
Class	Monoazo
Code numbers	CI (1975) No 19140 CI (1975) Food Yellow 4 CAS No 1934-21-0 EEC No E102
Chemical name	Trisodium 5-hydroxy-1-(4-sulfonatophenyl-4-(4-sulfonatophenylazo) pyrazole-3-carboxylate
Chemical formula	$C_{16}H_9N_4Na_3O_9S_2$
Structural formula	(see structure)
Molecular weight	534.37
Assay	Content not less than 85% total colouring matters
DESCRIPTION	Light orange powder or granules
FUNCTIONAL USE	Food colour

CHARACTERISTICS

IDENTIFICATION TESTS

**A. Solubility	Soluble in water Sparingly soluble in ethanol
**B. Identification of colouring matters	

---

* Supersedes the earlier specifications for Tartrazine published in FAO Nutrition Meetings Report Series No 38B (1966).

** See General Methods (Guide to Specifications, FAO Food and Nutrition Paper No 5, Revision 1, 1983).

PURITY TESTS

*Loss on drying at 135°	Not more than 15%
*Chloride and sulfate calculated as sodium salts	Not more than 15%
*Water insoluble matter	Not more than 0.2%
Subsidiary colouring matters	Not more than 1% (See description under TESTS)

Organic compounds other than colouring matters

> 4-hydrazinobenzene sulfonic acid
> 4-aminobenzene-1-sulfonic acid
> 5-oxo-1-(4-sulfophenyl)-2-pyrazoline-3-carboxylic acid
> 4,4'-diazoaminodi(benzene sulfonic acid)
> Tetrahydroxysuccinic acid

Total not more than 0.5% (See description under TESTS)

Unsulfonated primary aromatic amines:	Not more than 0.01% calculated as aniline
*Ether extractable matter	Not more than 0.2%
*Arsenic	Not more than 3 mg/kg
*Lead	Not more than 10 mg/kg
**Heavy metals	Not more than 40 mg/kg

TESTS

PURITY TESTS

*Subsidiary colouring matters

> Use the following conditions:
> Developing solvent: No. 4
> Height of ascent of solvent front: approximately 12 cm

*Organic compounds other than colouring matters

> Use HPLC under the following conditions:
> HPLC elution gradient: 2 to 100% at 2% per minute (exponential)

METHOD OF ASSAY

*Determination of Total Colouring Matters by Titration with Titanous Chloride

> Use the following:
> Weight of sample: 0.6-0.7 g
> Buffer: 15 g sodium hydrogen tartrate
> Weight (D) of colouring matters equivalent to 1.00 ml of 0.1 N $TiCl_3$: 0.01336 g

---

*   See Annex.

** See General Methods (Guide to Specifications, FAO Food and Nutrition Paper No 5, Revision 1, 1983).

TITANIUM DIOXIDE*
(Tentative)

DEFINITION	Titanium dioxide consists essentially of pure titanium dioxide which may be coated with small amounts of alumina and/or silica to improve the technological properties of the product.
Class	Inorganic colouring matter
Code numbers	CI (1975) No 77891 CI (1975) Pigment White 6 CAS No 13463-67-7 EEC No E171
Chemical name	Titanium dioxide
Chemical formula	TiO_2
Molecular weight	79.90
Assay	Content not less than 99% on an alumina and silica-free basis
DESCRIPTION	Amorphous white powder
FUNCTIONAL USE	Food colour

CHARACTERISTICS

IDENTIFICATION TESTS

**A. Solubility Insoluble in water and organic solvents.

**B. Solubility Dissolves slowly in hydrofluoric acid and in hot concentrated sulfuric acid.

C. Colour reaction Passes test
(See description under TESTS)

PURITY TESTS

<u>Aluminium oxide and silicon dioxide</u> Total not more than 2%
(See description under TESTS)

***<u>Loss on drying at 105°</u> Not more than 0.5%

<u>Loss on ignition at 800°</u> Not more than 1.0% on a volatile matter free basis
(See description under TESTS)

<u>Water soluble matter</u> Not more than 0.5%
(See description under TESTS)

---

* Supersedes the earlier specifications for Titanium Dioxide published in FAO Nutrition Meetings Report Series No 57 (1977).

** See General Methods (Guide to Specifications, FAO Food and Nutrition Paper No 5, Revision 1, 1983).

*** See Annex.

PURITY TESTS (cont'd)

<u>Matter soluble in 0.5 N HCl</u>	Not more than 0.5% on an alumina and silica-free basis and, in addition, for products containing alumina and/or silica, not more than 1.5% on the basis of the product as sold. (See description under TESTS)
<u>Arsenic</u>	Not more than 3 mg/kg by total dissolution. (See description under TESTS)
<u>Lead</u>	Not more than 10 mg/kg by total dissolution. (See description under TESTS)
<u>Mercury</u>	Not more than 1 mg/kg by total dissolution. (See description under TESTS)
<u>Antimony</u>	Not more than 50 mg/kg by total dissolution. (See description under TESTS)
<u>Zinc</u>	Not more than 50 mg/kg by total dissolution. (See description under TESTS)

TESTS

IDENTIFICATION TESTS

Colour reaction

Add 5 ml of sulfuric acid to 0.5 g of the sample, heat gently until fumes of sulfuric acid appear, then cool. Cautiously dilute to about 100 ml with water and filter. When a few drops of hydrogen peroxide are added to 5 ml of the clear filtrate an orange-red colour appears immediately.

PURITY TESTS

Aluminium oxide and/or silicon dioxide

Aluminium oxide

Reagents and sample solutions

0.01 N Zinc sulfate:
Dissolve 2.9 g of zinc sulfate ($ZnSO_4 \cdot 7H_2O$) in sufficient water to make 1000 ml. Standardize the solution as follows:

Dissolve 500 mg of high-purity (99.9%) aluminium wire, accurately weighed, in 20 ml of concentrated hydrochloric acid, heating gently to effect solution, then transfer into a 1000-ml volumetric flask, dilute to volume with water, and mix.

Transfer a 10 ml aliquot of this solution into a 500 ml Erlenmeyer flask containing 90 ml of water and 3 ml of concentrated hydrochloric acid, add 1 drop of methyl orange TS and 25 ml of 0.02 <u>M</u> disodium ethylenediaminetetraacetate (EDTA), and continue as directed below under <u>Sample Solution C</u>, beginning with "Add dropwise ammonia solution (1 in 5) until...".

0.01 N Zinc sulfate   (cont'd)

Calculate the titer T of the zinc sulfate solution by the formula:

$$T = \frac{18.896 \times W}{V}$$

in which: T is in terms of mg of Al_2O_3 per ml of zinc sulfate solution;
W is the weight of the aluminium wire taken in grams;
V is the volume, in ml, of the zinc sulfate solution consumed in the second titration;
18.896 is a factor derived as follows:

$$\frac{\text{mol.wt. } Al_2O_3}{\text{mol.wt. } Al} \times \frac{1000 \text{ mg}}{g} \times \frac{10 \text{ ml}}{2}$$

Sample Solution A:
Fuse 1 g of the sample, accurately weighed, with 10 g of sodium bisulfate ($NaHSO_4H_2O$) contained in a 250-ml high-silica glass Erlenmeyer flask. (*Caution:* Do not use more sodium bisulfate than specified, as an excess concentration of salt will interfere with the EDTA titration later on in the procedure.)

Begin heating at low heat on a hot plate, then gradually raise the temperature until full heat is reached. When spattering has stopped and light fumes of SO_3 appear, heat in the full flame of a Meker burner, with the flask tilted so that the fusion is concentrated at one end of the flask. Swirl constantly until the melt is clear (except for silica content), but guarding against prolonged heating to avoid precipitation of titanium dioxide. Cool, add 25 ml sulfuric acid solution (1 in 2) and then heat until the mass has dissolved and a clear solution results. Cool, and dilute to 120 ml with water.

Sample Solution B:
Measure out 200 ml of approximately 6.25 $\underline{M}$ sodium hydroxide, and add 65 ml of it to Sample Solution A while stirring constantly with a magnetic stirrer; pour the remaining 135 ml of the alkali solution into a 500-ml volumetric flask.

Slowly, and with constant stirring, add the sample mixture to the alkali solution in the 500-ml volumetric flask, then dilute to volume with water, and mix. (*Note:* If the procedure is delayed at this point for more than 2 hours, store the contents of the volumetric flask in a polyethylene bottle.)

Allow most of the precipitate to settle out (or centrifuge for 5 minutes), then filter the supernatant liquid through a very fine filter paper. Label the filtrate Sample Solution B.

Sample Solution C:
Transfer 100 ml of Sample Solution B into a 500-ml Erlenmeyer flask, add 1 drop of methyl orange TS, acidify with hydrochloric acid solution (1 in 2), and then add about 3 ml in excess. Add 25 ml of 0.02 $\underline{M}$ disodium ethylenediaminetetraacetate, and mix. [*Note:* If the approximate Al_2O_3 content is known, calculate the optimum volume of EDTA solution to be added by the formula (4 x % Al_2O_3) + 5.]

Add, dropwise, ammonia solution (1 in 5) until the colour is just completely changed from red to orange-yellow, then add 10 ml of ammonium acetate buffer solution (77 g of ammonium acetate plus 10 ml of glacial acetic acid, dilute to 1000 ml with water) and 10 ml of diammonium hydrogen phosphate solution (150 g of diammonium hydrogen phosphate in 700 ml of water, adjusted to pH 5.5 with a 1 in 2 solution of hydrochloric acid, then dilute to

./.

Sample Solution C (cont'd)

1000 ml with water). Boil for 5 minutes, cool quickly to room temperature in a stream of running water, add 3 drops of xylenol orange TS, and mix. If the solution is purple, yellow-brown, or pink, bring the pH to 5.3 - 5.7 by the addition of acetic acid; at the desired pH a pink colour indicates that not enough of the EDTA solution has been added, in which case another 100 ml of Sample Solution B should be taken and treated as directed from the beginning of the description of "Sample Solution C", except that 50 ml, rather than 25 ml, of 0.02 $\underline{M}$ disodium ethylenediaminetetraacetate should be used.

Procedure

Titrate Sample Solution C with 0.01 $\underline{N}$ zinc sulfate to the first yellow-brown or pink end-point colour that persists for 5-10 seconds.

(*Caution:* This titration should be performed quickly near the end-point by adding rapidly 0.2 ml increments of the titrant until the first colour change occurs; although the colour will fade in 5-10 seonds, it is the true end-point. Failure to observe the first colour change will result in an incorrect titration. The fading end-point does not occur at the second end-point. This first titration should require more than 8 ml of titrant, but for more accurate work a titration of 10-15 ml is desirable.)

Add 2 g of sodium fluoride, boil the mixture for 2-5 minutes, and cool in a stream of running water. Titrate the EDTA (which is released by fluoride from its aluminium complex) with 0.01 $\underline{N}$ zinc sulfate to the same fugitive yellow-brown or pink end-point as described above.

Calculation

Calculate the percentage of aluminium oxide (Al_2O_3) in the sample taken by the formula:

$$\frac{V \times T}{2 \times S}$$

in which: V is the volume, in ml, of 0.01 $\underline{N}$ zinc sulfate consumed in the second titration;
T is the titer of the zinc sulfate solution, determined previously;
S is the weight of the sample taken, in grams.

Silicon dioxide

Fuse 1 g of the sample, accurately weighed, with 10 g of sodium bisulfate ($NaHSO_4H_2O$) contained in a 250-ml high-silica glass Erlenmeyer flask. Heat gently over a Meker burner, while swirling the flask, until decomposition and fusion are complete and the melt is clear, except for the silica content, and then cool.

(*Caution:* Do not overheat the contents of the flask at the beginning, and heat cautiously during fusion to avoid spattering.)

To the cold melt add 25 ml of sulfuric acid solution (1 in 2), and heat very carefully and very slowly until the melt is dissolved. Cool, and carefully add 150 ml of water, pouring very small portions down the sides of the flask, with frequent swirling, to avoid overheating and spattering. Allow the contents of the flask to cool, and then filter through fine ashless filter paper, using a 60 degree gravity funnel. Wash out all of the silica from the flask onto the filter paper with sulfuric acid solution (1 in 10). Transfer the filter paper and its contents into a platinum crucible, dry in

Silicon dioxide (cont'd)

an oven at 120°, and then heat the partly covered crucible over a Bunsen burner. To prevent flaming of the filter paper, heat first the cover from above, and then the crucible from below.

When the filter paper is consumed, transfer the crucible to a muffle furnace and ignite at 1000° for 30 minutes. Cool in a desiccator, and weigh. Add 2 drops of sulfuric acid (1 in 2) and 5 ml of concentrated hydrofluoric acid (sp.gr. 1.15), and carefully evaporate to dryness, first on a low-heat hot plate (to remove the HF) and then over a Bunsen burner (to remove the H_2SO_4). Take precautions to avoid spattering, especially after removal of the HF. Ignite at 1000° for 10 minutes, cool in a desiccator, and weigh again. Record the difference between the two weights as the content of SiO_2 in the sample.

Loss on ignition

Accurately weigh 2 to 3 g of the sample in a platinum dish and ignite the sample at 800°C. Cool in a desiccator and reweigh. Calculate the percentage loss in weight. Subtract the percentage loss on drying at 105°C (previous test) and report the result as loss on ignition at 800°C on a volatile matter free basis.

Matter soluble in 0.5 N hydrochloric acid

Suspend 5 g of the sample in 100 ml of 0.5 N hydrochloric acid and heat on a steam bath for 30 minutes with occasional stirring. Filter through a Gooch crucible fitted with a glass fibre filter paper. Wash with three 10-ml portions of 0.5 N hydrochloric acid, evaporate the combined filtrate and washings to dryness and ignite at a dull red heat to constant weight.

Water soluble matter

Proceed as directed under Matter Soluble in 0.5 N HCl (above), using water in place of 0.5 N hydrochloric acid.

Arsenic, lead, mercury, antimony, zinc

(Method required)

METHOD OF ASSAY

Transfer about 300 mg of the sample, previously dried at 105° for 3 hours and accurately weighed, into a 250-ml beaker, add 20 ml of sulfuric acid and 7 to 8 g of ammonium sulfate, and mix. Heat on a hot plate until fumes of sulfuric acid appear, and continue heating over a strong flame until the sample dissolves or it is apparent that the undissolved residue is siliceous matter. Cool, cautiously dilute with 100 ml of water, and stir. Heat carefully to boiling while stirring, allow the insoluble matter to settle, and filter. Transfer the entire residue to the filter, and wash thoroughly with cold dilute sulfuric acid TS. Dilute the filtrate with water to 200 ml, and cautiously add about 10 ml of stronger ammonia TS to reduce the acid concentration to about 5% by volume of sulfuric acid.

Prepare a zinc amalgam column in a 25 cm Jones reductor tube, placing a pledget of glass wool in the bottom of the tube and filling the constricted portion of the tube with zinc amalgam prepared as follows:

Add 20 to 30 mesh zinc to a 2% mercuric chloride solution, using about 100 ml of the solution for each 100 g of zinc. After about 10 minutes, decant the solution from the zinc, then wash the zinc with water by decantation. Transfer the zinc amalgam to the reductor tube, and

./.

METHOD OF ASSAY (cont'd)

wash the column with 100-ml portions of dilute sulfuric acid TS until 100 ml of the washing does not decolourize 1 drop of 0.1 $\underline{N}$ potassium permanganate.

Place 50 ml of ferric ammonium sulfate TS in a 500-ml suction flask, and add 0.1 $\underline{N}$ potassium permanganate until a faint pink colour persists for 5 minutes. Attach the Jones reductor tube containing the zinc amalgam column to the neck of the flask, and pass 50 ml of dilute sulfuric acid TS through the tube at a rate of about 30 ml per minute. Pass the prepared titanium solution through the column at the same rate, followed by 100 ml each of dilute sulfuric acid TS and water. During these operations, keep the tube filled with solution or water above the upper level of the amalgam column. Gradually release the suction, wash down the outlet tube and the sides of the receiver, and titrate immediately with 0.1 $\underline{N}$ potassium permanganate. Perform a blank determination, substituting 200 ml of dilute sulfuric acid (1 in 20) for the sample solution, and make any necessary correction.

Each ml of 0.1 $\underline{N}$ potassium permanganate is equivalent to 7.990 mg of TiO_2.

XANTHOPHYLLS*
(Tentative)

SYNONYMS	Tagetes extract
DEFINITION	Xanthophylls are obtained by the solvent extraction of the petals or leaves of a number of different species of edible plants and also from *Tagetes erecta L* and alpha-alpha (Lucerne). The solvents used are **. After removal of the solvent the product may be mixed with edible vegetable oils. The main colouring principle is lutein but xanthophylls also contain other pigments and other substances such as oils, fats and waxes derived from the source material.
Class	Carotenoid
Code numbers	CAS No 127-40-2 (Lutein) EEC No E161b (Lutein)
Chemical name	Lutein is β, ε-carotene-3,3'-diol
Chemical formula	Lutein is $C_{40}H_{56}O_2$
Structural formula	Lutein has the following structure:

[Structural formula of lutein]

Molecular weight	Lutein: 568.85
Assay	Content not less than 1% total colouring matters
DESCRIPTION	
FUNCTIONAL USE	Food colour
CHARACTERISTICS	

IDENTIFICATION TESTS

***A. Solubility   Insoluble in water

---

* Supersedes the earlier specifications for Xanthophylls published in FAO Nutrition Meetings Report Series (1959) Volume II.

** Information required.

*** See General Methods (Guide to Specifications, FAO Food and Nutrition Paper No 5, Revision 1, 1983).

PURITY TESTS

    Residual solvents   (Information required)

  *Arsenic   Not more than 3 mg/kg

  *Lead   Not more than 10 mg/kg

**Heavy metals   Not more than 40 mg/kg

TESTS

METHOD OF ASSAY   (Information required)

The total colouring matters content is calculated as Lutein.

---

* See Annex.

** See General Methods (Guide to Specifications, FAO Food and Nutrition Paper No 5, Revision 1, 1983).

GENERAL SPECIFICATIONS FOR ALUMINIUM LAKES OF COLOURING MATTERS

DEFINITION   Aluminium lakes are prepared by reacting colouring matter complying with the purity criteria set out in the appropriate specification monograph with alumina under aqueous conditions. The alumina is usually freshly prepared undried material made by reacting aluminium sulfate or chloride with sodium carbonate or bicarbonate or ammonia. Following lake formation, the product is filtered, washed with water and dried.

Assay   Content of colouring matter shall be within the range specified by the vendor.

CHARACTERISTICS

IDENTIFICATION TESTS

*Solubility   Insoluble in water

**Identification of colouring matter

PURITY TESTS

**Water soluble chloride and sulfate calculated as sodium salts	Not more than 2%
**HCl insoluble matter	Not more than 0.5%
**Ether extractable matter	Not more than 0.2% (Method II)
**Arsenic	Not more than 3 mg/kg
**Lead	Not more than 10 mg/kg

---

* See General Methods (Guide to Specifications, FAO Food and Nutrition Paper No 5, Revision 1, 1983).

** See Annex.

LIST OF FOOD COLOURS FOR WHICH THE EXISTING SPECIFICATIONS HAVE BEEN WITHDRAWN

In reviewing the specifications for certain food colours, the Committee noted that these had been developed many years ago and that since that time no new information had been received. The Committee concluded that there was no apparent commercial utilization of 27 of these colours for food purposes. The Committee considered an additional 4 substances which had either no existing specifications and/or no ADI allocated. Specifications for the substances listed below were therefore withdrawn or not prepared.

In order to develop specifications for these substances the Committee requires information on methods of manufacture, chemical composition, and use in food.

- Acid fuchsine FB
- Alkanet and Alkannin
- Anthocyanine colour from Grape skin
- Anthocyanins
- Benzyl violet 4B
- Black 7984
- Blue VRS
- Brown FK
- Chocolate Brown FB
- Chrysoine
- Eosine
- Fast yellow AB
- Guinea green B
- Indanthrene blue RS
- Light green SF yellowish
- Methyl violet
- Naphthol yellow S
- Orange G
- Orange GGN
- Orange I
- Orange RN
- Orchil and Orcein
- Ponceau 2R
- Ponceau 6R
- Ponceau SX
- Quercitin and Quercitron
- Red 10B
- Rhodamine B
- Scarlet GN
- Sudan G
- Sudan Red G
- Ultramarines
- Violet 5 BN
- Yellow 2G
- Yellow 27175 N

ANNEX

METHODS OF ANALYSIS

ANNEX

Index

	Page
Identification of Colouring Matters	A-1
Determination of	
Total Colouring Matters Content	A-5
- by Spectrophotometry	A-5
- by Titration with Titanous Chloride	A-7
Subsidiary Colouring Matter	A-9
Organic Compounds Other than Colouring Matters	A-13
- by HPLC Principle	A-13
- by Column Chromatography	A-14
Unsulfonated Primary Aromatic Amines	A-17
Leuco Base in Sulfonated Triarylmethane Colouring Matters	A-19
Residual Solvents	A-21
Ether Extractable Matter	A-24
- Method I	A-24
- Method II	A-25
Matter Insoluble in Carbon Tetrachloride or Chloroform	A-26
Matter Insoluble in Water	A-27
Hydrochloric Acid-Insoluble Matter in Lakes	A-28
Metallic Impurities	A-29
- Instrumental Method for Determination of Arsenic, Lead, Copper, Antimony, Chromium, Zinc and Barium	A-29
- Determination of Antimony, Chromium, Copper, Lead and Zinc	A-32
- Determination of Barium	A-33
- Determination of Arsenic and Antimony	A-34
- Limit Test for Chromium	A-35
Mercury	
- Instrumental Method	A-37
- Colourimetric Method	A-41
- Limit Test	A-43
Water Content (Loss on Drying)	A-45
Chloride as Sodium Chloride	A-46
Sulfate as Sodium Sulfate	A-47
Sulfated Ash	A-48
Water Soluble Chlorides and Sulfates in Aluminium Lakes	A-49

IDENTIFICATION OF COLOURING MATTERS

Many of the colours used by food manufacturers are blends of colouring matters of the type described in the monographs, and some of the blends contain added diluent. A simple test to establish whether a powder sample is a single colouring matter or a physical blend of a number of colours is to sprinkle a very small quantity of the powder into each of two beakers, one containing water and the other containing concentrated sulfuric acid. Under these conditions the specks of individual colouring matters can easily be seen as they dissolve and the test is surprisingly sensitive.

The positive identification of individual food colours is often quite difficult. A large number are the sodium salts of sulfonic acids and this results in their having no precise melting point or boiling point. In addition synthesized colours usually contain subsidiary colouring matters while colouring matters extracted from natural sources generally contain a number of different colouring principles. Identification therefore is best achieved by comparison of the observed properties with the properties of authentic commercial samples. The principle techniques in use are chromatography and spectrophotometry and frequently both are required. For example, the presence of subsidiary colouring matters may so affect the observed spectra that positive identification of the principal component cannot be made. For this reason, it is advisable to separate the colouring matters by column, paper or thin layer chromatography before additional means of identification are attempted.

Paper and thin layer chromatography are often very useful in identification of colouring matters and do not require expensive equipment. But it must be kept in mind that the R_f-value of a substance is only theoretically a constant. In practice many factors, most of which are beyond control, have such an important influence that the R_f-values become a very unreliable quantity. These factors include: composition and age of the solvent mixture, concentration of solvent vapour in the atmosphere, quality of the paper, machine direction, kind and quality of subsidiary substances, concentration, pH-value of the solution, and temperature. For this reason, comparative chromatography should always be used. By simultaneous running of several substances of similar concentration a number of these factors are eliminated.

Coincidence of migration distances with a single solvent system should be looked upon only as one criterion of identity and further tests should be made for securing the finding.

The following table contains examples of the R_f-values that may be expected when 1% aqueous solutions of various colouring matters are subjected to thin layer chromatography on Silica Gel G in ten different solvent systems. The compositions of the solvent systems, all of which must be freshly prepared, are:

Solvent No.	
1.	iso-Propanol:ammonia (S.G. 0.880):water (7:2:1)
2.	iso-Butanol:ethanol:water:ammonia (S.G. 0.880) (10:20:10:1)
3.	Saturated aqueous potassium nitrate solution
4.	Phenol:water (4:1, w/v)
5.	Hydrochloric acid (S.G. 1.18):water (23:77)
6.	Trisodium citrate:ammonia (S.G. 0.880):water (2g:15ml:85ml)
7.	Acetone:ethylmethylketone:ammonia (S.G. 0.880):water (60:140:1:60)
8.	n-Butanol:ethanol:pyridine:water (2:1:1:2)
9.	iso-Propanol:ammonia (S.G. 0.880) (4:1)
10.	n-Butanol:acetic acid (glacial):water (10:5:6)

A-2

R_f values of some permitted and non-permitted water-soluble colours

	C.I. No.	EEC Serial No.	Solvent Number (see section 4)									
			1	2	3	4	5	6	7	8	9	10
REDS												
Ponceau 4R or Cochineal Red A	16255	E124	0.66 (0.85)	0.75	0.88	0.03	0.95	1.00	0.60	0.90	0.11	0.52 (0.00-0.57)
Carmoisine or Azorubine	14720	E122	0.65 (0.77)	0.81	0.00-0.42	0.16	0.00 (0.00-0.32)	1.00	0.65	0.88	0.34 (0.46)	0.63 (0.11-0.70)
Amaranth	16185	E123	0.62 (0.48,0.76)	0.75 (0.83)	1.00 (0.00-1.00)	0.04 (0.16)	0.00 (0.00-0.90)	1.00	0.40 (0.64,0.66)	0.90	0.10 (0.41)	0.39,0.67
Erythrosine BS	45430	E127	0.85 (0.68,0.79)	0.91 (0.86,0.74, 0.81)	0.00 (0.00-0.10)	0.00 (0.41)	0.00-0.71	0.00-0.95	0.64,0.66 (0.58)	0.89	0.66 (0.57,0.43)	1.00
Red 2G	18050	–	0.68	0.80	0.37	0.12	0.00-0.71	1.00	0.64	0.90	0.36	0.68
ORANGES												
Orange G	16230	–	0.71 (0.67,0.88)	0.80 (0.75)	0.64 (1.00,0.35)	0.23,0.15, 0.04	0.73	1.00	0.64 (0.62,0.50, 0.67)	0.91	0.36 (0.32,0.17)	0.69 (0.46,0.82)
Orange RN	15970	–	0.83 (0.62)	0.88 (0.78)	0.00 (0.00-0.42)	0.42 (0.13)	0.13 (0.38)	0.76 (1.00)	0.68 (0.65)	0.92	0.64 (0.29)	0.82,0.71
Sunset Yellow FCF or Orange Yellow S	15985	E110	0.75 (0.68)	0.82 (0.74)	1.00 (0.00-1.00)	0.17,0.03	1.00	1.00	0.65 (0.48)	0.90	0.34 (0.10,0.22)	0.67 (0.46)
YELLOWS												
Tartrazine	19140	E102	0.66	0.77	0.46,1.00	0.08	1.00	1.00	0.52 (0.62)	0.93	0.14 (0.21)	0.50
Yellow 2G	18965	–	0.63	0.80	0.77	0.21	0.74	1.00	0.62	0.92	0.21	0.75 (0.11-0.75)
Quinoline Yellow	47005	E104	0.83,0.88 (0.82)	0.88	0.00-1.00	0.65 (0.21)	0.26-1.00, 0.00-0.38	0.95 (0.35)	0.54 (0.68)	0.88	0.00-0.31 (0.83)	0.67 (0.46)
Fast Yellow AB	13015	E105	0.77	0.81	1.00	0.14	0.97	1.00	0.56	0.93	0.36	0.66
GREENS, BLUES AND VIOLETS												
Green S or Acid Brilliant-Green BS or Lissamine Green	44090	E142	0.44 (0.52,0.68, 0.74)	0.61 (0.67,0.75, 0.81,0.84)	0.49 (0.24)	0.53 (0.05,0.36, 1.00)	0.29 (0.43)	1.00	0.46 (0.56,0.71)	0.75 (0.89,0.92)	0.07	0.55
Indigo Carmine or Indigotin	73015	E132	0.56 (0.70)	0.50-0.76 (0.78)	0.00	0.09,0.18 (0.52)	0.92	0.94	0.66 (0.71,0.73)	0.89,0.84	0.37 (0.00-0.34)	0.00-0.63
Indanthrene Blue or Solanthrene Blue RS or Anthragen Blue	69800	E130	0.00	0.00	0.00 (1.00)	0.00	0.00	0.00	0.00	0.00	0.00	0.00
Brilliant Blue FCF	42090	–	0.64 (0.73)	0.78	0.05	0.45 (0.68)	0.10	0.00-1.00	0.61 (0.68)	0.88	0.30 (0.49, 0.00-0.10)	0.53 (0.64)
Patent Blue V	42051	E131	0.34-0.60	0.68	0.05	0.55	0.15	0.95	0.69 (0.72)	0.84 (0.92)	0.37,0.45, 0.70,0.76 (0.00-0.23)	0.64 (0.71)
Violet 6B	42640	–	0.73 (0.67,0.91)	0.80 (0.72)	0.00	0.62 (0.51,1.00)	0.00-0.37	0.00-1.00	0.00-0.68 (0.62)	0.89 (0.62)	0.94,0.87 (0.75-0.79)	0.00-0.75
Methyl Violet	42535	–	0.91 (0.80)	0.56,0.81 (0.90)	0.00 (0.00-0.31)	0.79-1.00	0.00-0.80	0.00 (0.00-0.53)	0.00,0.11,0.28, 0.53,1.00	0.11 (0.90)	0.00,0.10,0.87 (0.00-0.83)	0.00 (0.00-0.70)
BROWNS, BLACKS												
Brown FK	–	–	0.78,0.71, 0.66	0.79,0.86	1.00 (0.00-1.00)	0.69,0.27, 0.53(0.37), 0.15,0.00	0.00-0.77	1.00	0.59,0.64, 0.53(0.37), 0.36,0.51	0.93	0.34 (0.26,0.53)	0.00-0.73
Chocolate Brown FB	42640	–	0.00-0.69	0.00-0.75	0.00-1.00	0.00 (0.00-0.23)	0.00-1.00	1.00	0.62 (0.34,0.43, 0.62)	0.87	0.00-0.38	0.00-0.75
Chocolate Brown HT	20285	–	0.00-0.63	0.74	0.00-0.82	0.00 (0.00-0.16)	1.00	0.00-1.00	0.38	0.88	0.00-0.32	0.00-0.73
Black PN or Brilliant-Black BN	28440	E151	0.66 (0.47)	0.75	1.00 (0.00-1.00)	0.00	1.00	1.00	0.38 (0.61)	0.85	0.05	0.00-0.43
Black 7984	27755	E152	0.62	0.75	1.00	0.00	1.00	1.00	0.38 (0.61)	0.85	0.09	0.00-0.45

C.I. No.: Colour Index Number; ( ): Figures in parentheses indicate subsidiary spots of lower intensity;
0.x-0.y: Streak between the spots.

From Pearson, D. (1973) *J. Assoc. Public. Anal.*, 11, 137-138. Reprinted from Environmental Carcinogens - Selected Methods of Analysis, Vo. 4 - Some Aromatic Amines and Azo Dyes in the General and Industrial Environment (IARC Publications No.40), International Agency for Research on Cancer, Lyon, 1981.

Assessment of the colour shade should be made while the chromatograms are still moist with solvent and then again after drying. The shade should be assessed in incident and transmitted daylight as well as in UV light. In ultraviolet light many colouring matters show characteristic colour changes. Furthermore, it is thus often possible to trace colourless fluorescent impurities. If possible, two UV emitters which yield different wave lengths should be used; one lamp should emit around 250 nm.

Tests with acids, alkalis and other suitable reagents, in order further to safeguard the results, should be made. All tests may be carried out with fine capillary pipettes on each colour spot.

The following requirements should be met when identifying colouring matters:

- equal migration distances in several solvents;
- equal shade in daylight and ultraviolet light;
- equal colour changes with reagents.

Spectrophotometric methods of examination are among the most useful means of identification of colouring matters. Ultraviolet, visible, and infrared regions are all employed.

The visible region of the spectrum is ordinarily examined as the first step in attempting to identify an unknown colouring matter. Many colouring matters show characteristic absorption in the visible region while others do not. Spectra in the ultraviolet region may also be of use, and should be obtained if possible.

Infrared absorption spectra often offer the best means of identification of various compounds, but there are some difficulties in the practical use of the methods.

In the application of visible and ultraviolet spectrophotometry, spectra should always be obtained in more than one solvent, or if in a single solvent, under various conditions. Spectra of water solutions should be obtained under neutral (buffered with ammonium acetate) acid (0.1 $\underline{N}$ hydrochloric acid), and alkaline (0.1 $\underline{N}$ sodium hydroxide) conditions.

Absorption spectra are ordinarily plotted as charts, showing the absorbance at all wave lengths. Examination of the resulting curves should include more than location of the wave length of maximum absorption. The entire curve should be carefully inspected to determine the particular shape of the curve since "shoulders" or inflection points may be the most characteristic and useful features of the absorption spectra. These features often make it possible to distinguish between two or more colouring matters that have absorption maxima at the same wave length. Many colouring matters can be definitely characterized by observing the extent to which the absorption maxima and other features of the absorption curve are changed by variation in pH or by other changes in the solvent.

Infrared spectra can be obtained in several ways; the more commonly used are:

- spectra of solutions of the material in suitable solvents;
- spectra of suspensions of the material in a suitable liquid;
- the potassium bromide pellet technique
 (a small amount of the colouring matter, usually from 1 to 3 mg, is thoroughly mixed with pure, dry potassium bromide, the mixture is transferred to a suitable die and pressed into solid form by exerting a pressure of 700 to 1400 kg/square cm).

The spectra obtained are plotted in the usual way. Salient features of the resulting curves are the wave length at which absorption peaks occur, and the shape of the curves near the peaks.

Detailed discussion of the interpretation of infrared spectra is beyond the scope of this brief statement. It must be pointed out, however, that care must be taken to ensure that absorption peaks due to inorganic contaminants are not considered as due to the colouring matter.

The crystal structure or other physical state of the sample may affect the spectra obtained from suspensions or potassium bromide pellets. It is necessary to make certain that the unknown material has been treated in exactly the same manner as was the standard or known sample. Water soluble colouring matters can often be handled by dissolving the materials in water, adding a little acetic acid, evaporating to apparent dryness, and then drying at about 100° to remove the residual water. All materials to be tested should be free from water or other solvent before an infrared spectrum is obtained. Water and all organic solvents absorb infrared radiation.

As an example of the use of infrared spectra, two of the colouring matters, one of which is listed in this publication, may be mentioned. Sunset Yellow and Orange GGN have absorption spectra in the visible region and in the ultraviolet region so nearly identical that they cannot be distinguished through examination in these regions. Their infrared spectra, however, are quite different in the region of the spectrum in which the sulfonic acid groups absorb strongly.

In some instances, chromatographic procedures, spectrophotometric procedures, and any combination of the two may fail to give positive identification. In such cases, the problem may often be solved by reducing the colouring matter or otherwise degrading the compound and identifying the resulting products. This technique is particularly applicable to azo colouring matters. The amino compounds resulting from the reduction can frequently be readily identified by chromographic and spectrophotometric techniques.

Many other techniques have been applied to identification of colouring matters. Their description here will not be undertaken, but one example will be cited. Many pigments have definite crystalline structure and can be identified by X-ray diffraction patterns, or by optical crystallography. Some colouring matters can be converted to crystalline derivatives and similarly identified.

DETERMINATION OF TOTAL COLOURING MATTERS CONTENT

Two general methods are used for determination of total colouring matters content:

Spectrophotometric comparison
and
Titanous chloride reduction

When using the spectrophotometric method it should be remembered that many makers of spectrophotometers do not guarantee accuracy greater than $\pm$ 1%. All colours present in the sample which have their absorption peak in the region of that of the main colour will contribute to the absorbance figure used to calculate the results. Subsidiary colouring matters of markedly different hue will not be accounted for by this technique. In the ideal situation the spectrophotometer would be used to make side-by-side comparison of the sample and a standard of known colouring content. The use of a generally accepted absorptivity figure in place of the standard itself is considered to be somewhat less satisfactory, but in the case of a number of colours it is recognized that there is no practical alternative.

In the titanous chloride method the assumption is made that isomers and subsidiaries have the same titanous chloride equivalent as the main colouring matter.

Determination of Total Colouring Matters by Spectrophotometry

Two experimental procedures are described. They differ only in detail, and in both the calculation makes use of the absorptivity figure quoted in the colour specification.

The first is typical of the type used for water soluble colouring matters; the second is suitable for solvent soluble colouring matters, expecially the synthesized carotenoids. (The solutions prepared in the second procedure are used in the identification tests for the carotenoids.)

Principle

The absorbance of a solution of the colouring matter is determined at its wavelength of maximum absorption and compared with a standard absorptivity figure.

Apparatus

- Spectrophotometer capable of accurate ($\pm$ 1% or better) measurement of absorbance in the region of 350-700 nm with an effective slit width of 10 nm or less.
- Absorption cells, 1 cm light path.

Reagents

Freshly distilled water *or* the solvent prescribed in the specification for the colour.

./.

Determination of Total Colouring Matters by Spectrophotometry (cont'd)

Procedure 1

Accurately weigh 0.25 g (± 0.02 g) of the sample. Transfer to a 1000 ml volumetric flask. Add freshly distilled water *or* the prescribed solvent and swirl to dissolve. Make up to volume and mix. Dilute to a solution of suitable strength according to the details given in the colour specification. Determine the absorbance (A) at its wavelength of maximum absorption in a 1 cm cell.

Calculation

Calculate the total colouring matters content of the sample using the following equation:

$$\text{Total colouring matters} = \frac{A}{a} \times \frac{vol}{wt} \times 100$$

where A = absorbance of the sample
a = absorptivity of the standard (from the colour specification)
vol = dilution factor = $\frac{\text{Volume diluted to}}{\text{Volume measured}}$
wt = weight of sample taken

Procedure 2

Accurately weigh about 0.08 g (=w) of sample in a 100 ml volumetric flask (=V_1) and dissolve by shaking briefly with 20 ml pure, acid-free chloroform. Make sure that the solution is clear. Make up to volume by the addition of pure cyclohexane. Pipet 5.0 ml of the solution (=v_1) into a 100 ml volumetric flask (=V_2) and make up to volume with cyclohexane.

Similarly, dilute 5.0 ml of this solution (=v_2) to 100 ml and measure the absorbance at the absorbance maximum (=A) against cyclohexane as a blank, using 1 cm cells. Calculate the specific absorbance according to the following formula:

$$\%\text{ total colouring matters} = \frac{A \cdot V_1 \cdot V_2}{v_1 \cdot v_2 \cdot w \cdot a}$$

where A = absorbance of the sample solution at the wavelength of maximum absorption
V_1 = volume of first volumetric flask (=100 ml)
V_2 = volume of second volumetric flask (=100 ml)
v_1 = volume of first pipet (=5 ml)
v_2 = volume of second pipet (=5 ml)
w = mass of sample in g
a = absorptivity

Complete the determination promptly, avoiding exposure to air insofar as possible, and undertaking all operations in the absence of daylight.

Procedure for Lakes

Procedure 1, above, can be adapted in the following manner for the determination of the total colouring matter content of lakes.

Determination of Total Colouring Matters by Spectrophotometry (cont'd)

Procedure for Lakes (cont'd)

Prepare pH 7 phosphate buffer in the following way:

Dissolve 13.61 g of potassium dihydrogen phosphate in water and dilute to 1000 ml. Add about 90 ml of $\underline{N}$ sodium hydroxide. Determine the pH using a pH-meter and make any fine adjustment of the pH to 7.0 using approximately 0.1 $\underline{N}$ sodium hydroxide or diluted phosphoric acid.

Accurately weigh a quantity of lake which will give an absorbance approximately equal to that of the parent colour when the latter is tested according to Procedure 1, above. Transfer to a beaker containing 10 ml concentrated hydrochloric acid diluted to approximately 50 ml with water. Heat with stirring to dissolve the lake, then cool to ambient temperature and dilute to exactly 1000 ml with pH 7 phosphate buffer. Then proceed as detailed in Procedure 1, above, and in the colour monograph, using pH 7 phosphate buffer as the solvent.

Determination of Total Colouring Matters Content by Titration with Titanous Chloride

Apparatus

- Carbon dioxide generator, e.g. a Kipp apparatus.
- Titration apparatus, as shown in figure, comprising an aspirator and a burette fitted with a double-oblique tap and side arm. The aspirator is closed at the top by a rubber stopper with two holes, through one of which passes a glass tube connected to the carbon dioxide generator. Through the second hole passes a glass tube connected to the top of the burette. The side arm of the burette is connected to the bottom of the aspirator.
- 500 ml conical flask with CO_2 inlet tube. (When carrying out tirations with titanous chloride, ensure that the tip of the burette is always inside the neck of the flask.)

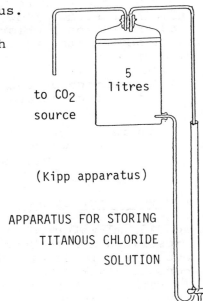

(Kipp apparatus)

APPARATUS FOR STORING TITANOUS CHLORIDE SOLUTION

Procedure

Preparation of 0.1 $\underline{N}$ titanous chloride solution.
Measure into a large round flask a volume of 15% titanous chloride solution containing 15.5 - 16.5 g of titanous chloride for each litre of solution required. Add 100 ml of hydrochloric acid (sp.gr. 1.16 - 1.18) for each litre of solution required. Boil the mixture for 1-2 minutes and then pour it into cold water. Make up the required volume in the aspirator, mix well, and preserve the solution in an atmosphere of carbon dioxide. Cover the solution with a layer of liquid paraffin (about 0.6 cm deep).

./.

Determination of Total Colouring Matters Content
by Titration with Titanous Chloride (cont'd)

Procedure (cont'd)

Standardization of titanous chloride solution.
Weigh 3 g of ammonium ferrous sulfate [$(NH_4)_2SO_4 \cdot FeSO_4 \cdot 6H_2O$] into a 500 ml conical flask, and pass a stream of carbon dioxide through the flask continuously until the end of the determination. Add 50 ml of water and 25 ml of 10 $\underline{N}$ sulfuric acid solution, then 30 ml of 0.1 $\underline{N}$ potassium dichromate solution, accurately standardized. Titrate with the titanous chloride solution until the calculated end point is nearly reached. Add 5 ml of a 20% ammonium thiocyanate solution and continue the titration until the red colour is discharged and the solution remains green. Carry out a blank determination on 3 g of ammonium ferrous sulfate, using the same quantities of water, acid and ammonium thiocyanate solution and passing a continuous current of carbon dioxide through the flask as before.

The factor for 0.1 $\underline{N}$ titanous chloride solution is:

$$F = \frac{30}{\text{ml (corrected) of titanous chloride solution required}}$$

Determination of total colouring matter content of sample.
Accurately weigh the quantity of the colour sample specified in each monograph into a 500 ml flask and add 10 g of sodium citrate or 15 g of sodium hydrogen tartrate, as specified in each monograph, and 150 ml of water. Pass a stream of carbon dioxide through the flask, heat the solution to boiling, and titrate with standardized titanous chloride solution, maintaining a steady flow of carbon dioxide. Towards the end of the titration, add the titanous chloride solution dropwise to the end point. The colour will act as its own indicator unless otherwise stated in the appropriate monograph.

$$\text{Percentage of total colouring matters in sample} = \frac{A \times D \times 100 \times F}{\text{weight of sample}}$$

where A = ml of 0.1 $\underline{N}$ titanous chloride solution required (corrected)
      D = weight of colouring matters equivalent to 1.00 ml of 0.1 $\underline{N}$ titanous chloride solution (specified in each monograph)

Procedure for Lakes

Add 150 ml water to the 500 ml conical flask and dissolve in it the buffer prescribed for the parent colour. Accurately weigh a quantity of lake equivalent to 35 - 40 ml of 0.1 $\underline{N}$ titanous chloride and transfer it to the conical flask. Heat to boiling or until the lake has completely dissolved. Titrate with titanous chloride solution in the manner described under Procedure, above.

DETERMINATION OF SUBSIDIARY COLOURING MATTER

General Note

For many years paper chromatography has been used for determining the subsidiary colouring matter content of water-soluble food colours. In the commonly used version of this well-established technique the assumption is made that the absorptivities of subsidiary colouring matters are similar to that of the main colouring matter. Accordingly, standards of individual subsidiary colour matters are not required.

HPLC has been used successfully to separate and determine the subsidiary colouring matter contents of a number of food colours, including some of the water soluble ones. Standards of individual subsidiary colouring matters are needed for this method. However, it should be remembered that specification limit figures are, unless otherwise stated, linked to the paper chromatographic method and the conditions referred to under "Tests".

Definition

Subsidiary colouring matters are defined as those colouring matters that are produced during the manufacturing process in addition to the principal named colouring matter(s). Any colouring matters other than the principal and subsidiary colouring matters are considered to be adulterants and their presence is usually detected on the chromatograms used to determine subsidiary colouring matters. Interpretation of the chromatograms for adulterant colours usually requires some experience.

Principle

The subsidiary colouring matters are separated from the main colouring matter by ascending paper chromatography and are extracted separately from the paper. The absorbance of each extract is measured at its wavelength of maximum absorption in the visible spectrum.

Because it is impractical to identify each subsidiary colouring matter in each food colour, an approximate method of expressing them as a percentage of the sample has been adopted. The assumption is made that the absorptivity of each subsidiary colouring matter is the same as that of the total colouring matters. The absorbances of the extracts are added together and used in conjunction with the absorbance of the sample and its total colouring matters content to calculate the subsidiary colouring matters content. This is considered to be a sufficiently close approximation for the determination of a minor component.

Apparatus

- Chromatography tank and ancillary equipment.
  Suitable apparatus is shown in Figs. 1 and 2 and comprises:
    - a glass tank (A) and cover (B);
    - a supporting frame (C) for the chromatography paper sheets;
    - a tray (D) for chromatography solvent;
    - a secondary frame (E) supporting "drapes" of the filter paper;
    - sheets of chromatography grade paper not less than 20 cm x 20 cm (Whatman No.1 Chromatography grade paper is suitable).
- Microsyringe, capable of delivering 0.1 ml with a tolerance of ± 0.002 ml
- Spectrophotometer.

Procedure

Not less than 2 hours before carrying out the determination, arrange the filter-paper drapes in the glass tank and pour over the drapes and into the bottom of the tank sufficient of the chromatography solvent to cover the bottom of the tank to a depth of approximately 1 cm. Place the solvent tray (D) in position and fit the cover to the tank.

Mark out a sheet of chromatography paper as shown in Fig. 3. Apply 0.10 ml of a 1.0% aqueous solution of the sample as uniformly as possible within the confines of the 18 cm x 7 cm rectangle, holding the nozzle of the microsyringe steadily in contact with the paper. Allow the paper to dry at room temperature for 1 - 2 hours or at $50°$ for 5 minutes, followed by 15 minutes at room temperature. Mount the sheet, together with a plain sheet to act as a blank, in frame (C).

Pour sufficient of the chromatography solvent into the tray (D) to bring the surface of the solvent about 1 cm below the base line of the chromatogram sheet. The volume necessary will depend on the dimensions of the apparatus and should be predetermined. Put frame (C) into position and replace the cover. Allow the solvent front to ascend the specified distance above the base line, then remove frame (C) and transfer it to a drying cabinet at $50° - 60°$ for 10 - 15 minutes. Remove the sheets from frame (C).

(If required, several chromatographs may be developed simultaneously.)

Cut each subsidiary band from the sheet as a strip, and cut an equivalent strip from the corresponding position of the plain sheet. Place each strip, subdivided into a suitable number of approximately equal portions, in a separate test tube. Add 5.0 ml of water:acetone (1:1 by vol.) to each test tube, swirl for 2 - 3 minutes, add 15.0 ml of 0.05 $\underline{N}$ sodium hydrogen carbonate solution and shake the tube to ensure mixing. Filter the coloured extracts and blanks through 9 cm filter papers of open texture and determine the absorbances of the coloured extracts at their wavelengths of maximum absorption, using 40 mm closed cells, against a filtered mixture of 5.0 ml of water:acetone (1:1 by vol.) and 15.0 ml of the 0.05 $\underline{N}$ sodium hydrogen carbonate solution. Measure the absorbances of the extracts of the blank strips at the wavelengths at which those of the corresponding coloured extracts were measured.

To prepare the "standard", proceed as follows:

> From the 1.0% solution of the sample prepare a solution corresponding to L/100% where L = the subsidiary colouring matters limit. Apply 0.10 ml of this solution to a sheet of chromatography paper by the technique outlined above and then dry it at $50° - 60°$ for 10 - 15 minutes. Cut the band from the sheet as a strip and cut an equivalent strip from a plain but marked sheet. Proceed as detailed previously and determine the net absorbance (A_S) of the standard.

$$\% \text{ subsidiary dyes} = \frac{a + b + c \ldots n}{A_S} \times L \times \frac{D}{100}$$

> where a, b and c etc. = the net absorbances of the subsidiary colouring matters; and
> D = the total colouring matters content of the sample.

Chromatography solvents

1. Water:ammonia (S.G.O.880):trisodium citrate (95 ml:5 ml:2 g)

2. n-Butanol:water:ethanol:ammonia (S.G.O.880) (600:264:135:6)

3. Butan-2-one:acetone:water (7:3:3)

4. Butan-2-one:acetone:water:ammonia (S.G.O.880) (700:300:300:2)

5. Butan-2-one:acetone:water:ammonia (S.G.O.880) (700:160:300:2)

6. n-Butanol:glacial acetic acid:water (4:1:5)
   Shake for 2 minutes, allow layers to separate. Use the upper layer as the chromatography solvent.

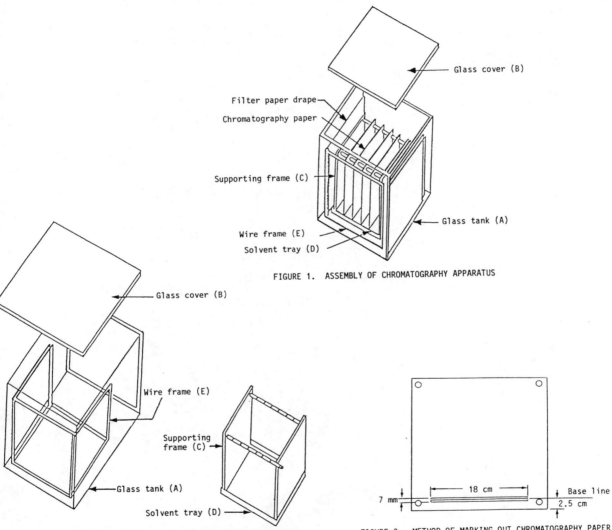

FIGURE 1. ASSEMBLY OF CHROMATOGRAPHY APPARATUS

FIGURE 2. COMPONENTS OF CHROMATOGRAPHY APPARATUS

FIGURE 3. METHOD OF MARKING OUT CHROMATOGRAPHY PAPER

DETERMINATION OF ORGANIC COMPOUNDS OTHER THAN COLOURING MATTERS

General Note

For the separation and determination of uncoloured impurities, High Performance Liquid Chromatography (HPLC) has several advantages over other chromatographic techniques, viz. improved separations, speed (it can be automated) and accuracy. When determining named organic compounds a sample of each material likely to be encountered is needed before any particular colour can be analysed. It is usual for HPLC methods to be outlined rather than described in detail.

It should be remembered that HPLC is still in a period of considerable development. Column packing materials, capillary columns, type and sensitivity of detectors are some of the aspects which are continuing to receive attention from the manufacturers and in due course their development may lead to the separation of impurities in addition to those currently listed in some food colour specifications.

The method selected uses an SAX column but reverse phase columns are widely used and give excellent separations.

The alternative (traditional) method uses column chromatography, the technique being to collect the eluant in fractions and use their ultra-violet absorption spectra to identify the compounds present and to calculate their concentrations.

DETERMINATION OF ORGANIC COMPOUNDS OTHER THAN COLOURING MATTERS BY HPLC PRINCIPLE

The organic compounds other than colouring matters are separated by HPLC using gradient elution and quantitatively determined by comparison of their peak areas against those obtained from standards. The conditions prescribed must be treated only as guidelines and minor modifications may be needed to achieve the separations.

The conditions used in performing the chromatography are given below:

Instrument – High Performance Liquid Chromatograph fitted with a gradient elution accessory.

Detector – A UV HPLC detector recording absorbances at 254 nm.

Column – Stainless steel, 1 m x 2.1 mm internal diameter.

Column Packing – Pellicular strong anion exchange (SAX), e.g., quaternary ammonium substituted methacrylate polymer coated 1% by weight.

Sample Concentration – 2% weight/weight in 0.01M sodium tetraborate.

Injection Volume – 10 μl.

Solvent System – Primary: 0.01 M sodium tetraborate.
Secondary: 0.01 M sodium tetraborate/0.1 M sodium perchlorate.

Gradient – See individual monograph.

Flow Rate – 1.0 ml per minute.

Temperature – Ambient.

Alternative experimental conditions such as column length, types of column packing and solvent system, and the use of paired ion procedures, may produce variations in elution characteristics such as order of elution and resolution

DETERMINATION OF ORGANIC COMPOUNDS OTHER THAN COLOURING MATTERS BY COLUMN CHROMATOGRAPHY

Apparatus

Chromatographic tube (see Figure). Suitable spectrophotometer for use in the ultraviolet range.

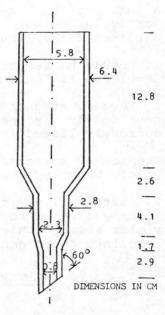

CHROMATOGRAPHIC TUBE

Column Preparation

Prepare a slurry of Whatman powdered cellulose in a 25% ammonium sulfate (very low in iron) solution. If other cellulose is used, the iron content must be very low.

> Test: Prepare column as directed and pass 200 ml of 25% ammonium sulfate solution through it. The ultraviolet absorption of the solution must be sufficiently low to avoid interference with the intended analysis. Use about 75 g of cellulose to 500 ml of liquid. Place a small disk of stainless steel gauze in the constriction above the tip of the tube. Pour sufficient volume of the flurry into the tube to give a column to a height of about 5 cm in the mouth of the tube. Tap the tube occasionally to ensure a well-packed column. Wash the column with 200 ml of the eluant.

Procedure

Place 0.200 g of the colour sample in a suitable beaker and dissolve in 20 ml of water. Add approximately 5 g of powdered cellulose. Add 50 g of ammonium sulfate to salt out the colour. Transfer the mixture to the chromatograph tube, rinse the beaker with the 25% ammonium sulfate solution and add the washings to the tube. Allow the column to drain until flow ceases, or nearly so.

Add the ammonium sulfate solution to the column at a rate equivalent to the rate of the flow through the column. Collect the effluent in 100 ml fractions. Continue until twelve fractions have been collected. Reserve the column and contents until the last fraction has been examined.

Mix each fraction well, and obtain the ultraviolet absorption spectra of each solution from 220 to 400 nm. If the spectrum of the twelfth fraction shows the presence of any compound, continue collecting fractions until the compounds present are eluted.

Usually only one compound is encountered. Identification and quantitative determination is accomplished by comparison of the absorption spectra of the eluted material with the spectra of solutions of the pure compounds in the same solvent.

When more than one compound is present in significant quantities in any fractions, the spectrophotometric data will so indicate. In such case, the amounts of the various compounds must be determined by the procedure customarily used in spectrophotometric analysis of mixtures of absorbing materials.

./.

Procedure (cont'd)

Some samples contain small amounts of various materials, particularly inorganic salts, that contribute "background absorption". Correction for this is made as follows:

> Determine the amount of such absorption of the fraction collected from the column immediately before and of the fraction immediately following those fractions in which the compounds are encountered. Subtract one-half of the sum of these two determinations from the observed absorbance of the fractions containing the compounds. The remainder is taken as the absorbance due to the compound present.

DETERMINATION OF UNSULFONATED PRIMARY AROMATIC AMINES

Principle

Unsulfonated primary aromatic amines are extracted into toluene from an alkaline solution of the sample, re-extracted into acid and then determined spectrophotometrically after diazotisation and coupling. They are expressed as aniline unless they are known to be some other amine.

Apparatus

- Visible range Spectrophotometer

Reagents

The reagents shall be of a recognized analytical reagent quality. Distilled water or water of at least equal purity shall be used.

1. Toluene
2. Hydrochloric acid, approx N solution
3. Hydrochloric acid, approx $3N$ solution
4. Potassium bromide, approx 50% solution
5. Sodium carbonate, approx $2N$ solution
6. Sodium hydroxide, approx N solution
7. Sodium hydroxide, approx $0.1\ N$ solution
8. R salt (2-naphthol-3:6-disulfonic acid, disodium salt), approx. $0.05\ N$ solution
9. Sodium nitrite, approx $0.5\ N$ solution
10. Standard aniline solution:

    Into a small weighing beaker weigh 0.100 g of redistilled aniline, then wash it into a 100-ml one-mark volumetric flask, rinsing the beaker several times with water. Add 30 ml of approx 3 N hydrochloric acid solution and dilute to the mark with water at room temperature. Call this Solution A.

    Dilute 10.0 ml of Solution A with water to 100 ml in a one-mark volumetric flask and mix well. Call this Solution B; 1 ml of this solution will be equivalent to 0.00001g of aniline. (Prepare Solution B freshly when required.)

Procedure

Preparation of Calibration Graph

Measure the following volumes of standard aniline Solution B into a series of 100 ml one-mark volumetric flasks: 5 ml, 10 ml, 15 ml, 20 ml, 25 ml.

Dilute to 100 ml with approx N hydrochloric acid solution and mix well. Pipet 10 ml of each mixture into clean, dry test tubes and cool for 10 minutes by immersion in a beaker of ice/water mixture. To each tube add 1 ml of the potassium bromide solution and 0.05 ml of the sodium nitrite solution. Mix and allow to stand for 10 minutes in the ice/water bath. Into each of five 25 ml volumetric flasks, measure 1 ml of the R salt solution, and 10 ml of the sodium carbonate solution. Pour each diazotised aniline solution into a separate flask containing R salt solution, rinsing the test tubes with a few drops of water. Dilute to the mark with water, stopper the flasks, mix the contents well and allow to stand for 15 minutes in the dark.

./.

Procedure (cont'd)

Preparation of Calibration Graph (cont'd)

Measure the absorbance of each coupled solution at 510 nm in 40 mm cells, using as a reference a mixture of 10.0 ml of N hydrochloric acid solution, 10.0 ml of the sodium carbonate solution, and 2.0 ml of the R salt solution, diluted to 25.0 ml with water. Plot a graph relating optical density to weight of aniline in each 100 ml of aniline solution.

Preparation and Examination of Test Solution

Weigh, to the nearest 0.01 g, about 2.0 g of the colour sample into a separating funnel containing 100 ml of water, swill down the sides of the funnel with a further 50 ml of water, swirl to dissolve the sample, and add 5 ml of N sodium hydroxide solution. Extract with two 50-ml portions of toluene and wash the combined toluene extracts with 10-ml portions of 0.1 N sodium hydroxide solution to remove traces of colour. Extract the washed toluene with three 10-ml portions of 3 N hydrochloric acid solution and dilute the combined extract to 100 ml with water. Mix well. Call this Solution T.

Pipet 10.0 ml of Solution T into a clean, dry test tube, cool for 10 minutes by immersion in a beaker of ice/water mixture, add 1 ml of the potassium bromide solution and proceed as described above for the preparation of the calibration graph, starting with the addition of 0.05 ml of the sodium nitrite solution.

Use as the reference solution in the measurement of absorbance a solution prepared from 10.0 ml of the test solution, 10 ml of the sodium carbonate solution and 2.0 ml of the solution, diluted to 25.0 ml with water.

Read from the calibration graph the weight of aniline corresponding to the observed optical density of the test solution.

Calculation

Percentage of unsulfonated primary aromatic amine (as aniline) in sample =

$$\frac{\text{weight of aniline} \times 100}{\text{weight of sample taken}}$$

DETERMINATION OF LEUCO BASE IN
SULFONATED TRIARYLMETHANE COLOURING MATTERS

Principle

Air is blown through an aqueous solution containing the chloride and dimethylformamide. Under these conditions the leuco base is oxidized to colouring matters and the increase in absorptivity is a measure of the amount of leuco base originally present.

Reagents

Dimethylformamide (DMF)

Solution A: Weigh 10.0g of $CuCl_2 \cdot 2H_2O$ and dissolve in 200 ml of DMF: Transfer to a 1-litre volumetric flask and make up to the mark with DMF.

Solution B: Accurately weigh the specified quantity of sample. Dissolve in approximately 100 ml water, transfer quantitatively to a 1-litre volumetric flask and make up to the mark with water.

Procedure

Prepare the following solutions:

Solution a: Pipet 50 ml DMF into a 250-ml volumetric flask. Cover with parafilm and place in the dark.

Solution b: Accurately pipet 10 ml of Solution B into a 250-ml volumetric flask. Add 50 ml DMF. Cover with parafilm and place in the dark.

Solution c: Pipet 50 ml of Solution A into a 250-ml volumetric flask. Bubble air through this solution for 30 minutes in the following manner:

Insert a 5 ml pipette into a box attached to a bench air flow source. Turn on the air, slowly. Stick the pipette down into the solution in the flask and adjust the air flow to a rapid but controlled rate. After 30 minutes pull the pipette up out of the solution and rinse the sides of the pipette into the flask with H_2O from a wash bottle. Then turn off the air flow.

Solution d: Accurately pipet 10 ml Solution B into 2 separate 250-ml volumetric flasks in the same manner as used for Solution b. Add 50 ml Solution A to each flask. Bubble air through the solutions for 30 minutes, using the above method.

After 30 minutes of rapid bubbling of air through the solutions, dilute all 5 flasks nearly to volume with water. Heat is evolved when DMF and water are mixed, so place the flasks in a water bath of tap water until they have cooled to room temperature. Do not leave them for longer than necessary; 5 - 10 minutes is normally long enough. Bring accurately to volume with water. Run the solutions on the spectrophotometer immediately. The entire procedure should be completed as quickly as possible.

./.

Spectrophotometric Determination

Draw the following curves from 700 - 500 nm using an absorbance range of 0.1 and 1 cm cells. Run all curves on the same spectrogram, and (for maximum accuracy) take readings off the numerical display at the maximum between 620 and 635 nm by cranking back after the curve is drawn.

Curve	Reference Cell	Sample Cell	Comments
1	a	a	Set zero at 700 nm, run curve; record absorbance at Abs std for colouring matter
2	a	b	Run curve without readjusting zero setting; record absorbance at maximum
3	c	c	Set zero at 700 nm; record absorbance at Abs std for colouring matter
4a	c	d_1	Run curve without readjusting zero setting; record absorbance at maximum
4b	c	d_2	Run curve without readjusting zero setting; record absorbance at maximum

(d_1 and d_2 are duplicate determinations)

NOTE: Cells <u>must</u> be thoroughly rinsed before each run. For the flow-through cell, use 3 separate rinses of at least 40 ml of the sample solution to be run.

Calculations

$$\frac{[(4 - 3) - (2 - 1)] \times 25 \times 100\%}{a \times 1 \text{ cm} \times \text{mg sample} \times \text{ratio}} = \% \text{ leuco base}$$

where a = absorptivity of 100% colouring matters
mg sample = amount sample weighed in mg

$$\text{ratio} = \frac{\text{MW of colouring matter}}{\text{MW of leuco base}}$$

DETERMINATION OF RESIDUAL SOLVENTS

Apparatus

- Distilling Head. Use a Clevenger trap designed for use with oils heavier than water. A suitable design is shown in the Figure.

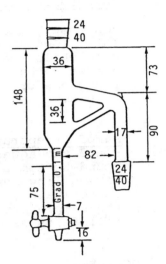

CLEVENGER TRAP
(Dimensions in mm)

Reagents and Solutions

Toluene
The toluene used for this analysis should not contain any of the solvents determined by this method. The purity may be determined by gas chromatographic analysis, using one of the following columns or its equivalent:

(1) 17% by weight of Ucon 75-H-90,000 on 35/80-mesh Chromosorb W;
(2) 20% Ucon LB-135 on 35/80-mesh Chromosorb W;
(3) 15% Ucon LB-1715 on 60/80-mesh Chromosorb W; or
(4) Porapak Q 50/60-mesh.

Follow the conditions described under Procedure, and inject the same amount of Toluene as will be injected in the analysis of the solvents. If impurities interfering with the test are present, they will appear as peaks occurring before the Toluene peak and should be removed by fractional distillation.

Benzene
The benzene used for this analysis should be free from interfering impurities. The purity may be determined as described under Toluene.

Detergent and Antifoam
Any such products that are free from volatile compounds may be used. If volatile compounds are present, they may be removed by prolonged boiling of the aqueous solutions of the products.

Reference Solution A
Prepare a solution in Toluene containing 2500 mg/kg of benzene. If the Toluene available contains benzene as the only impurity, the benzene level can be determined by gas chromatography and sufficient benzene added to bring the level to 2500 mg/kg.

Reference Solution B
Prepare a solution containing 0.63% v/w of acetone in water.

Sample Preparation A (all solvents except methanol)
Place 50.0 g of the sample, 1.0 ml of Reference Solution A, 10 g of anhydrous sodium sulfate, 50 ml of water, and a small amount each of Detergent and Antifoam in a 250-ml round-bottom flask with a 24/40 ground glass neck. Attach the Distilling Head, a 400-mm water-cooled

./.

Reagents and Solutions (cont'd)

Sample Preparation A (cont'd)

condenser, and a receiver, and collect approximately 15 ml of distillate. Add 15 g of anhydrous potassium carbonate to the distillate, cool while shaking, and allow the phases to separate. All the solvents except methanol will be present in the toluene layer, which is used in the Procedure. Draw off the aqueous layer for use in Sample Preparation B.

Sample Preparation B (methanol only)

Place the aqueous layer obtained from Sample Preparation A in a 50-ml round-bottom distilling flask with a 24/40 ground-glass neck, add a few boiling chips and 1.0 ml of Reference Solution B, and collect approximately 1 ml of distillate, which will contain any methanol from the sample, together with acetone as the internal standard. The distillate is used in the Procedure.

Procedure

Use a gas chromatograph equipped with a hot-wire detector and a suitable sample-injection system or on-column injection. Under typical conditions, the instrument contains a 63 mm (od) x 183 to 244 cm column maintained isothermally at $70°$ to $80°$. The flow rate of dry carrier gas is 50 to 80 ml per minute and the sample size is 15 to 20 μl (for the hot-wire detector).

The column selected for use in the chromatograph depends on the components to be analyzed and, to a certain extent, on the preference of the analyst. The columns 1, 2, 3 and 4, as described under Toluene, may be used as follows:

(1) This column separates acetone and methanol from their aqueous solution. It may be used for the separation and analysis of hexane, acetone, and trichloroethylene in the toluene layer from Sample Preparation A. The elution order is acetone, methanol, and water, or hexane, acetone, propan-2-ol plus dichloromethane, benzene and trichloroethylene plus toluene.

(2) This column separates dichloromethane and isopropanol. The elution order is hexane plus acetone, dichloromethane, propan-2-ol benzene, trichloroethylene, and toluene.

(3) This is the best general purpose column, except for the determination of methanol. The eultion order is hexane, acetone, benzene, and toluene.

(4) This column is used for the determination of methanol, which elutes just after the large water peak.

Calibration

Determine the response of the detector for known ratios of solvents by injecting known mixtures of solvents and benzene in toluene. The levels of the solvents and benzene in tolene should be of the same magnitude as they will be present in the sample under analysis.

./.

Calculation

Calculate the areas of the solvents with respect to benzene, and then calculate the calibration factor, F, as follows:

$$F \text{ (solvent)} = \frac{\text{weight solvent}}{\text{weight benzene}} \times \frac{\text{area of benzene}}{\text{area of solvent}}$$

The recovery of the various solvents from the sample, with respect to the recovery of benzene, is as follows:

> hexane – 52%
> acetone – 85%
> propan-2-ol – 100%
> dichloromethane – 87.5%
> trichloroethylene – 113%
> methanol – 87%

Calculate the mg/kg of residual solvent (except methanol) by the formula:

$$\text{Residual solvent} = \frac{43.4 \times F(\text{solvent}) \times 100}{\% \text{ recovery of solvent}} \times \frac{\text{area of solvent}}{\text{area of benzene}}$$

in which 43.4 is the mg/kg of benzene internal standard, related to the 50 g sample taken for analysis. Calculate the mg/kg of residual methanol by the formula:

$$\text{Residual methanol} = \frac{100 \, F}{0.87} \times \frac{\text{area of methanol}}{\text{area of acetone}}$$

in which 100 is the mg/kg of acetone internal standard, related to the 50 g sample taken for analysis, and F is the calibration factor determined by using known mixture of methanol and acetone.

DETERMINATION OF ETHER EXTRACTABLE MATTER

METHOD I

Apparatus

Upward displacement type liquid/liquid extractor with sintered glass distributor, working capacity 200 ml. A piece of bright copper wire is suspended through the condenser and a small coil of copper wire (0.5 g) is placed in the distillation flask.

Ether

Diethylether or diisopropylether.

Immediately before use the freshly distilled ether should be passed through a 30 cm column of chromatography grade aluminium oxide in order to remove peroxides and inhibitors. Test to ensure the absence of peroxides, as follows:

> Prepare a colourless solution of ferrous thiocyanate by mixing equal volumes of 0.1 N solutions of ferrous sulfate and ammonium thiocyanate and carefully discharging any red colouration, due to ferric ions, with titanous chloride. To 50 ml of this solution add 10 ml of ether and shake the mixture vigorously for 2-3 minutes. No red colour should develop.

Procedure

(i) Alkaline ether extract. Weigh accurately about 5.0 g [1]/ of the colour sample, dissolve in 150 ml of water, add 2 ml of 2.5 N sodium hydroxide solution and transfer the solution to the extractor, diluting to approximately 200 ml with water in the process. Add 200 ml of ether to the distillation flask and extract for 2 hours with a reflux rate of about 15 ml/min. Reserve the colour solution. Transfer the ether extract to a separatory funnel and wash the ether extract with two 25-ml portions of 0.1 N sodium hydroxide and then with water. Distil the ether in portions from a tared 150-ml flask containing a clean copper coil, reducing the volume to about 5 ml.

(ii) Acid ether extract. To the colour solution reserved from (i), add 5 ml of 3 N hydrochloric acid, mix and extract with a further quantity of the ether as in (i). Wash the ether extract with two 25-ml portions of 0.1 N hydrochloric acid and then with water. Transfer in portions to the flask containing the evaporated alkaline extract and carefully evaporate all the ether. Complete the drying in an oven at 85° for 20 minutes, then allow the flask to cool in a desiccator for 30 minutes and weigh. Repeat the drying and cooling until constant weight is obtained. The increase in weight of the tared flask, expressed as a percentage of the weight of sample taken, is the "ether extract".

1/ Some colours have solubilities of less than 5 g/150 ml and the use of a lower weight is prescribed in the Colour Specification under TESTS.

DETERMINATION OF ETHER EXTRACTABLE MATTER

METHOD II

Apparatus

Soxhlet extractor. A piece of bright copper wire is suspended through the condenser and a small coil of copper wire (0.5 g) is placed in the distillation flask.

Ether

Diethylether <u>or</u> diisopropylether.

Purify the ether as direction in METHOD I.

Procedure

Weigh accurately about 2 g of the colour sample. Transfer to the Soxhlet thimble and extract with 150 ml ether for 5 hours. Concentrate the ether extract on a steam bath to about 5 ml. Dry up the residue in a tared evaporating dish on a water bath and then dry at 105° until a constant weight is obtained.

The increase in weight of the evaporating dish, expressed as a percentage of the weight of sample taken, is the "ether extract".

DETERMINATION OF MATTER INSOLUBLE IN
CARBON TETRACHLORIDE OR CHLOROFORM

Apparatus

- Oven, 0 - 200° range
- Hot plate, capable of boiling carbon tetrachloride (CCl_4) (b.p. 76.8°)
- Gooch crucible, fitted with glass fiber disk
- Vacuum flask
- Source of vacuum
- Desiccator

Reagents

- Carbon tetrachloride, reagent grade, or
- Chloroform, reagent grade.

(These are referred to as "Solvent" in the "Procedure".)

Procedure

The test shall be carried out in accordance with the following instructions

Mix the prescribed weight of the sample (W_1) with 100 ml of solvent in a 250-ml beaker, stir and heat to boiling on the hot plate in a fume hood. Filter the hot solution through a weighed gooch crucible (W_2). Transfer residue in the beaker to the crucible with solvent. Wash the residue in the crucible with 10-ml portions of solvent until washings are colourless. Place the crucible in oven at 100 - 150° for 3 hours; cool crucible in desiccator. Weigh cooled cruicible (W_3).

Calculation

Calculate the percent carbon tetrachloride or chloroform insoluble matter (PIM) in the sample using the following equation:

$$PIM = \frac{W_3 - W_2}{W_1} \times 100$$

Report as percent carbon tetrachloride or chloroform insoluble matter in the original sample.

DETERMINATION OF MATTER INSOLUBLE IN WATER

Weigh 4.5 - 5.5 g [1] of the sample into a 250 ml beaker. Add about 200 ml of hot water (80-90°), stir to dissolve, and allow the solution to cool to room temperature. Filter the solution through a tared Grade 4 sintered glass filter (B.S. 1752, sintered disk filters for laboratory use) and wash with cold water until the washings are colourless. Dry the filter and residue at 135° until a constant weight is obtained. Express the weight of the residue as a percentage of the weight of sample taken.

---

[1] *NOTE:* Some colours have solubilities of less than 5 g/200 ml and the use of a lower weight is prescribed in the Monograph under "Tests".

DETERMINATION OF HYDROCHLORIC ACID-INSOLUBLE MATTER IN LAKES

Reagents

- Concentrated hydrochloric acid
- Hydrochloric acid 0.5% v/v

Procedure

Accurately weigh approximately 5 g of the lake into a 500 ml beaker. Add 250 ml water and 60 ml concentrated hydrochloric acid. Boil until all the colour and alumina has dissolved. Filter through a tared No.4 sintered glass crucible. Wash the crucible with hot 0.5% hydrochloric acid until the washings are colourless. Dry the crucible to constant weight at $135°$.

Express the weight of reside as a percentage of the weight taken.

DETERMINATION OF METALLIC IMPURITIES

All the procedures for trace metals commence with the desctruction of organic matter in the colour sample. The trace metal content may then be determined by instrumental or chemical methods.

Atomic spectroscopy combines speed with accuracy and is widely used when large numbers of samples have to be analysed.

Chemical methods depend on the formation of specific coloured compounds by the metal impurities. The colour intensities of sample and standards are then compared visually or by using a spectrophotometer.

General precautions

Because of the minute amounts of metals involved special care must be taken to reduce the reagent blanks to as low a value as possible and to avoid contamination during the test. All apparatus should be thoroughly cleaned with a mixture of hot dilute acids (1 part hydrochloric acid, 1 part concentrated nitric acid, and 3 parts water) followed by thorough washing with water immediately before use. The methods of preparation described should be followed exactly.

INSTRUMENTAL METHOD FOR DETERMINATION OF ARSENIC, LEAD, COPPER, ANTIMONY, CHROMIUM, ZINC AND BARIUM

Principle

The samples are digested in a mixture of sulfuric, nitric and perchloric acids. The lead, copper, chromium and zinc in the digest are determined by conventional flame atomic absorption spectroscopy. Antimony and arsenic are determined by using a hydride generation technique and barium is determined using atomic emission. Alternately, antimony may be determined by atomic absorption but the hydride generation technique is more sensitive.

Apparatus

- Kjeldahl flasks, of silica or borosilicate glass (nominal capacity 100 ml) fitted with an extension to the neck by means of a B24 ground joint, as shown in Figure 1. The extension serves to condense the fumes and carries a tap funnel through which the reagents are introduced.

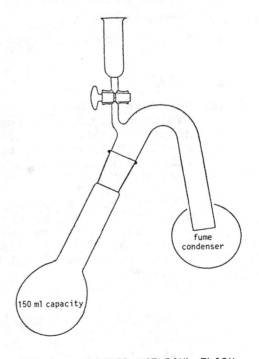

FIG. 1. MODIFIED KJELDAHL FLASK
(open type)

Apparatus (cont'd)

- Atomic absorption spectrophotometer.
  Any commercial instrument operating in both the absorption and emission mode may be used providing it has facilities for the selection of the required oxidant/fuel combination from a choice of air, argon, nitrous oxide, hydrogen and acetylene; has a wavelength range from 180 to 600 nm, and an emission scanning facility.

  A hydride generation vessel accessory is also required and is available from all the major commercial manufacturers of atomic absorption equipment. For operations in emission mode and measurements of absorption involving the generation of a gaseous hydride, a potentiometric recorder is necessary, preferably a multi-range type covering the range 1-20 mV.

Reagents

Reagents shall be of an order of purity higher than accepted analytical reagent grade quality. Metal-free water (see below) shall be used throughout.

(a) Nitric acid, sp.gr. 1.42

(b) Perchloric acid, 60% (w/w) solution

(c) Sulfuric acid, 98% H_2SO_4

(d) Hydrochloric acid, sp.gr. 1.16 – 1.18

(e) Hydrochloric acid 5 $\underline{N}$ solution prepared by dilution of reagent (d) with metal-free distilled water

(f) Water, metal free. Distilled water may be re-distilled from an all-glass apparatus or may be passed down a column of cation exchange resin, e.g., Amberlite IR 120 (H).

(g) Sodium sulfate

(h) Sodium borohydride pellets

Standards

(a) <u>Standard copper solution</u>
    Dissolve 3.928 g of pure copper sulfate $CuSO_4 \cdot 5H_2O$ in distilled water, dilute to 1000 ml at 20° with distilled water in a one-mark graduated flask. Dilute 10 ml to 100 ml with water in a one-mark graduated flask as required.
    1 ml = 100 μg Cu.

(b) <u>Standard zinc solution</u>
    Dissolve 1.000 g of pure zinc powder in a mixture of 10 ml distilled water and 5 ml hydrochloric acid [special reagent (d)] and dilute to 1000 ml at 20° with distilled water, in a one-mark graduated flask. Dilute 10 ml to 100 ml with water in a one-mark graduated flask as required.
    1 ml = 100 μg Zn.

(c) <u>Standard chromium solution</u>
    Dilute 5.80 ml of 0.1 $\underline{N}$ Potassium dichromate solution to 100 ml at 20° with distilled water in a one-mark graduated flask as required.
    1 ml = 100 μg Cr.

./.

Standards (cont'd)

(d) **Standard antimony solution**
Dissolve 2.668 g potassium antimony tartrate $K(SbO)C_4H_4O_6$ in distilled water, dilute to 1000 ml at 20° with distilled water in a one-mark graduated flask. Dilute 10.0 ml to 100 ml with distilled water in a one-mark graduated flask as required.
1 ml = 100 μg Sb.

(e) **Standard lead solution**
Dissolve 1.60 g of lead nitrate, $Pb(NO_3)_2$ in nitric acid (10 ml of concentrated nitric acid diluted with 20 ml water, boiled to remove nitrous fumes, and cooled) and dilute to 1000 ml with water in a one-mark graduated flask. Dilute 10.0 ml of this solution to 500 ml at 20° with water in a one-mark graduated flask as required.
1 ml = 20 μg Pb.

(f) **Standard barium solution**
Dissolve 1.779 g barium chloride $BaCl_2 2H_2O$ in distilled water, dilute to 1000 ml at 20° with distilled water in a one-mark graduated flask. Dilute 10.0 ml to 100 ml with distilled water in a one-mark graduated flask as required.
1 ml = 100 μg Ba.

(g) **Standard arsenic solution**
Dissolve 1.320 g of arsenious oxide, As_2O_3 by warming at a temperature not exceeding 60° with 14 ml of 5 $\underline{N}$ sodium hydroxide solution in a 100-ml beaker. Cool, add 0.2 ml of phenol phthalein indicator and neutralize with 6 $\underline{N}$ sulfuric acid. Transfer the solution to a 1000 ml one-mark graduated flask containing 10 g of sodium hydrogen carbonate dissolved in water, washing out the beaker with water. Dilute to the mark with water at 20° and mix. Dilute 5 ml of this solution to 1000 ml at 20° with water, in a one-mark graduated flask as required.
1 ml = 5 μg As.

Preparation of test solutions

Accurately weigh about 2.5 g of the colour sample into a 100 - 150 ml Kjeldahl flask, and add 5 ml of the dilute nitric acid. As soon as any initial reaction subsides, heat gently until further vigorous reactions cease and then cool. Add gradually 4 ml of concentrated sulfuric acid at such a rate as not to cause excessive frothing on heating (5-10 minutes are usually required) and then heat until the liquid darkens appreciably in colour, i.e., begins to char.

Add concentrated nitric acid slowly in small portions, heating between additions until darking again takes place. Do not heat so strongly that charring is excessive, or loss of arsenic may occur; a small but not excessive amount of free nitric acid should be present throughout. Continue this treatment until the solution is only pale yellow in colour and fails to darken in colour on prolonged heating. Now run in 0.5 ml of the perchloric acid solution and a little concentrated nitric acid and heat for about 15 minutes, then add a further 0.5 ml of the perchloric acid solution and heat for a few minutes longer. Note the total amount of concentrated nitric acid used. Allow to cool somewhat and dilute with 10 ml of water. The solution should be quite colourless (if much iron is present it may be faintly yellow). Boil down gently, taking care to avoid bumping, until white fumes appear. Allow to cool, add a further 5 ml of water and again boil down gently to fuming. Finally, cool, add 10 ml 5 $\underline{N}$ hydrochloric acid and boil gently for a few minutes. Cool and transfer

Preparation of test solutions (cont'd)

the solution to a 50-ml one-mark graduated flask washing out the Kjeldahl flask with small portions of water. Add the washings to the graduated flask and dilute to the mark with water. Call this Solution A.

Prepare a reagent blank using the same quantities of reagents as used in the sample oxidation.

DETERMINATION OF ANTIMONY, CHROMIUM, COPPER, LEAD AND ZINC

Preparation of Calibration Curve Solutions

To a series of 100-ml one-mark volumetric flasks pipet 0, 1, 2, 3, 4 and 5 of the appropriate standard solution [Standards (a) to (e)] and dilute to about 50 ml. Add 8 ml concentrated sulfuric acid [reagent (c)], 10 ml concentrated hydrochloric acid [reagent (d)] and 1.0 g sodium sulfate. Shake to dissolve and when solution is complete, dilute to the mark with metal-free water.

These solutions then contain 0, 1.0, 2.0, 3.0, 4.0 and 5.0 μg per ml of either copper, zinc, chromium or antimony, or 0, 0.2, 0.4, 0.6, 0.8 and 1.0 μg per ml of lead.

Instrumental Conditions

Select the wavelength and gases to be used for the particular element under consideration from the table below.

Element	Wavelength (nm)	Gases
Antimony	217.6	Air/acetylene
Chromium	357.9	Nitrous oxide/acetylene
Copper	324.7	Air/acetylene
Lead	283.3	Air/acetylene
Zinc	213.9	Air/acetylene

The recommended settings for the various instrumental parameters vary from model to model, and certain parameters require optimization at the time of use to obtain the best results. Instruments should therefore be adjusted as described in the manufacturer's instructions using the type of flame and wavelength settings specified above.

Procedure

Set the atomic absorption spectrophotometer to the appropriate conditions. Aspirate the strongest standard containing the element to be determined and optimize the instrument settings to give full-scale or maximum deflection on the chart recorder. Measure the absorbances of the other standards and plot a graph showing the net absorbance against the concentration of the element in the standard solutions. Aspirate the solution A obtained from the wet oxidation of the sample and the corresponding blank solution and determine the net absorbance. Using the graph prepared above, determine the concentration of the element in the sample solution.

$$\frac{\text{Concentration of element } (\mu g/ml) \times 50}{\text{Weight of sample taken (g)}} = \text{ppm element}$$

DETERMINATION OF BARIUM

The determination of low levels of barium by flame spectrometry necessitates the use of the hotter fuel rich nitrous oxide/acetylene flame with the addition of an easily ionized species such as sodium cations to act as ionization suppressant. Measurements are normally made in the more sensitive flame emission mode using a scanning technique rather than recording at a fixed wavelength. The recommended instrument settings are intended as a guide since variations in scan rate, speed of response, etc. vary considerably between individual instruments. The use of a potentiometric recorder is essential for recording the atomic emission spectra.

Preparation of calibration curve solution

Transfer into a series of 100-ml graduated flasks 0, 1, 2, 3, 4 and 5 ml aliquots of standard barium solution [Standard (f)] and dilute to approximately 50 ml with water. Carefully add 8 ml concentrated sulfuric acid [reagent (c)], 10 ml hydrochloric acid [reagent (d)] and 1.0 g sodium sulfate [reagent (g)]. Shake carefully to dissolve, allow to cool to ambient temperature and dilute to volume with water.

The standard solutions contain barium equivalent to 0, 1, 2, 3, 4 and 5 μg/ml Ba.

Instrumental conditions

- Scanning range — 550 – 560 bn (λ max Ba 553.5 nm) at 4 nm minute^{-1}
- Slit width — 20 Å units
- Recorder range — 20 mV
- Chart speed — 60 mm hr^{-1}
- Sensitivity — Variable on most commercial instruments, but should be adjusted so that the highest standard gives approximately three quarters full scale deflection. Sufficient sensitivity should be obtained using a standard nitrous oxide/acetylene head used for atomic absorption measurements.

Procedure

With instrument settings above, spray the series of freshly prepared standards into the flame for sufficient time to complete each wavelength scan (ca. 2-1/2 minutes). To reduce slit blockage during relatively long scans it may be necessary to spray distilled water for a short period between standards and samples.

Directly follow the standard scans by the sample solutions and if necessary repeat at least one of the standards every third sample to check for changes in response per unit weight barium.

Draw in the background emission on the recorder traces and measure the height of the peak above that of the base line. Plot a graph relating peak height to standard concentration and use the relationship to determine the weight of barium in the sample solutions. The graph should be linear over the range 0 – 5 μg ml^{-1} Ba.

Calculation: $$\frac{\mu g \ Ba \ ml^{-1} \times 50}{\text{Weight sample (g)}} = \text{ppm Barium}$$

DETERMINATION OF ARSENIC AND ANTIMONY

Arsenic and antimony are determined after preparation of their volatile hydrides which are collected either in the generation vessel itself or, in some designs, in a rubber balloon attached to the vessel. The gases are then expelled with Argon into a hydrogen flame.

Preparation of calibration curve solution

Into a series of 100-ml one-mark volumetric flasks add from a burette, 0, 1, 2, 3, 4 and 5 ml of standard arsenic or antimony solution [Standards (g) and (d)] and dilute to about 50 ml with distilled water. Add 8 ml concentrated sulfuric acid [reagent (c)], 10 ml hydrochloric acid [reagent (d)] and 1.0 g sodium sulfate [reagent (g)]. Shake to dissolve, and when solution is complete, dilute to the mark with distilled water.

Instrumental conditions

Using the atomic absorption spectrophotometer with the appropriate hollow cathode or electrodeless discharge lamp, select the wavelength for either arsenic (193.7 nm) or antimony (217.6 nm).

Procedure

Measure 5.0 ml of the strongest standard into the generation vessel, add 25 ml water and 2 ml 5 $\underline{N}$ hydrochloric acid [reagent (e)]. Stopper the vessel and expel any air as described in the maker's instructions, filling the apparatus with argon. Isolate the vessel from the atomizer using the by-pass valve. Remove the atomizer and then quickly add 1 pellet of sodium borohydride weighing approximately 0.2 g [reagent (h)] and replace the stopper. Ensure that all the joints are secure.

When the reaction slows (20 - 30 seconds) open the appropriate taps to allow argon to drive the generated hydride into the flame. When the hydride has all been expelled as shown by the recorder trace, return the taps to their original position and empty the vessel.

Optimize the instrument settings to give full scale deflection for the strongest standard. Measure the other standards, the sample and the blank solution using the same procedure.

Plot a graph relating peak height on the recorder to concentration of the arsenic or antimony in the standards. Using the net absorbance of the sample, read from the graph the concentration of arsenic or antimony in the solution.

Calculation:

$$\frac{\text{Concentration of arsenic or antimony } (\mu g/ml) \times 50}{\text{Mass of sample taken (g)}} = \text{ppm arsenic or antimony}$$

LIMIT TEST FOR CHROMIUM

The limit test described hereunder is designed to show whether the sample contains more or less than 20 mg/kg of chromium.

Procedure

Weigh 1.0 g of the sample into a quartz dish. Char the material, raising the temperature slowly. Allow to cool, add 10 ml of a 25% magnesium nitrate solution; evaporate, heating slowly until no more nitrous vapour evolves. Heat the material in an oven at 600° until all black particles have disappeared (1 hour).

Dissolve the residue by adding 10 ml of 4 $\underline{N}$ sulfuric acid and 20 ml of water. Heat on a water bath for about 5 minutes.

Add 0.5 ml of 0.1 $\underline{N}$ potassium permanganate. Add more permanganate if the solution decolourizes. Cover with a watch glass and heat on a water bath for about 20 minutes. Add 5% sodium azide solution, one drop every 10 seconds, until the excess potassium permanganate has been removed (avoid excess of sodium azide; 2 drops are usually sufficient). Cool the solution in running water, and filter if manganese dioxide is evident. Transfer the solution to a 50-ml volumetric flask. Add 2.5 ml of 5 $\underline{M}$ sodium dihydrogenphosphate, add 2 ml of diphenyl carbazide TS and fill to the mark with water. Measure the extinction at 540 mu 30 minutes after adding the diphenyl carbazide TS. A blank with the latter two reagents should show no colour or only a slight purple colour.

At the same time run a parallel test with 1.00 ml of standard chromate TS (1 ml = 20 μg Cr) and a few ml of saccharose placed into a second quartz dish. Treat the mixture exactly as the dye sample and measure the extinction at the same wave length.

Calculate the chromium content of the sample from the two extinction values observed.

INSTRUMENTAL METHOD FOR DETERMINATION OF MERCURY

General Remarks

The sample is ashed by heating under reflux with sulfuric and nitric acids. The oxidation is completed by addition of potassium permanganate solution. After successive additions of hydroxylamine hydrochloride solution and stannous chloride solution, the mercury content is measured by cold vapour atomic absorption spectrometry.

Special Reagents

(a) Nitric Acid, density 1.40 g/ml.

(b) Sulfuric acid, density 1.84 g/ml.

(c) Sulfuric acid, approximately 3.5 mol/l. Prepare by diluting 1 volume of concentrated sulfuric acid (b), with 4 volumes of water.

(d) Sulfuric acid, approximately 1 mol/l. Prepare by diluting 1 volume of 3.5 mol/l sulfuric acid (c) with 2.5 volumes of water.

(e) Hydrochloric acid, density 1.18 g/ml.

(f) Potassium permanganate solution, 50.0 g/l.

(g) Hydroxylamine hydrochloride solution, 100.0 g/l.

(h) Stannous chloride solution. Prepare by dissolving 25.0 g of stannous chloride ($SnCl_2 \cdot 2H_2O$) in 50 ml hydrochloric acid (e). Make up to 250 ml with water and bubble nitrogen through the solution. Store over a few granules of metallic tin.

(i) Chromic acid mixture. Dissolve 4.0 g of potassium dichromate in 300 ml of 3.5 mol/l sulfuric acid (c) and make up to 1 litre with water.

(j) Magnesium perchlorate, in granular form for gas desiccation.

(k) Mercuric chloride

Standards

(a) Mercuric Chloride Solution, 0.5 mg Hg/ml.
Weigh out, to the nearest 0.1 mg, 0.677 g of mercuric chloride (k). Dissolve in approximately 250 ml 3.5 mol/l sulfuric acid (c) in a 1-litre volumetric flask, add approximately 700 ml water and then potassium permanganate solution (f) dropwise until a colouration persists. Make up to the mark with water and mix well. Renew this solution every three months.

(b) Mercuric Chloride Solution, 0.02 µg Hg/ml.
Dilute the standard mercuric chloride solution 0.5 mg Hg/ml [Standard (A)] by a factor of 25,000 by successive dilution with sulfuric acid [special reagent (d)], e.g. 10 ml made up to 250 ml twice followed by 10 ml made up to 400 ml. Before bringing up to the mark in the final dilution, add potassium permanganate solution [special reagent (f)] dropwise until a colouration persists. Renew this solution daily.

./.

Apparatus

All the glassware must be cleaned with hot nitric acid [special reagent (a)] and washed thoroughly with water before use.

- <u>Mineralization apparatus</u> fitted with reflux condenser (see figure).

- <u>Bubblers</u>, with a ground glass stopper fitted with two tubes to permit entrainment of the mercury vapour and with a calibration mark at the required volume for measurement.

  The capacity of the bubbler and position of the mark depend on the atomic absorption spectrophotometer used.

  Clean the bubbler successively with chromic acid mixture [special reagent (i)], tap water and double distilled water before use.

- <u>Water vapour absorption apparatus</u>, containing magnesium perchlorate [special reagent (j)].

- <u>Atomic absorption spectrophotometer</u> suitable for the cold vapour determination of mercury in open or closed circuit, with recorder.

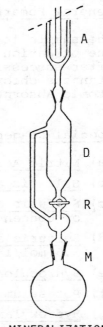

MINERALIZATION APPARATUS

Procedure

1. <u>Ashing</u>

   Weigh out, to the nearest 2 mg, approximately 0.5 g sample containing not more than 0.5 µg total mercury. Introduce the sample into the receiver flask (M), and add a few glass beads. Connect the receiver flask to the condensate reservoir (D) and close the stopcock (R).

   Introduce into the reservoir 25 ml of nitric acid [special reagent (a)] followed by 10 ml sulfuric acid [special reagent (b)]. Mount and turn on the condenser (A). Open the stopcock carefully and allow small portions of the mixture of acids to run into the receiver flask. Interrupt the flow of acids if the reaction becomes too vigorous.

   Empty the reservoir into the receiver flask, mix the contents of the latter well by careful shaking and leave the stopcock open.

   Heat the receiver flask carefully. As soon as foaming has ceased, close the stopcock (R), continue heating and let the condensate collect in the reservoir.

   Discontinue heating when the contents of the receiver flask begin to char. Allow a small portion of the condensate to run into the receiver flask, close the stopcock again and resume heating the receiver flask. Repeat this procedure for as long as the contents display charring when heated.

   When charring has ceased, heat and add condensate as soon as white fumes appear. Continue alternately heating and adding condensate for one hour. Finally, heat the contents of the flask to white fumes.

./.

Procedure (cont'd)

1. Ashing (cont'd)

Stop heating and allow to cool to approximately 40°. Open the stopcock and allow all the condensate to run into the receiver flask. Wash the apparatus out from the top of the condenser with 5 - 10 ml of water, collect the washings in the receiver flask and disconnect it from the reservoir.

2. Treatment of the Solution

Introduce the potassium permanganate solution [special reagent (f)] dropwise into the receiver flask, with agitation, until a pink colouration persists. Note the quantity of reagent (f) used. (If this quantity exceeds 10 ml, repeat the procedure "1. Ashing" as above.)

Heat gently to boiling, then allow to cool.

Pour the contents of the receiver flask into a bubbler, wash the receiver flask with water and add the washings to the contents of the bubbler.

Measure the mercury content (see below) the same day as the treatment of the solution.

3. Measurement of Mercury Content

Introduce 5 ml of hydroxylamine hydrochloride [special reagent (g)] into the bubbler and make up to the mark either with double distilled water or with sulfuric acid [special reagent (d)] in the case of standard solutions. Add 5 ml of stannous chloride solution [special reagent (g)], assemble the bubbler, connect it to the water vapour absorption apparatus and to the atomic absorption spectrophotometer. Set the latter in operation.

Mix the contents of the bubbler well by gentle shaking, pass air or nitrogen through, measure and record. Carry out measurements as quickly as possible after the addition of stannous chloride. If an open-circuit system is used, wait 30 seconds before passing air or nitrogen.

4. Calibration Curve

Introduce respectively 2, 5, 10, 15 and 25 ml aliquots of the standard mercury solution [Standard (b)] into bubblers and 25 ml sulfuric acid [special reagent (d)] into a sixth bubbler. Add potassium permanganate solution [special reagent (f)] dropwise, with agitation, to each bubbler until a colouration persists.

Measure the mercury content as described above.

Plot the calibration curve with the measured absorption values as ordinates and the corresponding mercury contents in micrograms as abscissae. The working standards contain 0.04, 0.10, 0.20, 0.30, 0.50 and 0 μg of mercury, respectively.

./.

Procedure (cont'd)

5. Method of Addition

   The method of addition may be used if an open-circuit system is used.

   Place one of the working standard solutions (see 4 above) in a bubbler and add an aliquot portion of the sample solution obtained after treatment (see 2 above). The quantity of mercury in the bubbler must lie in the range in which the photometer gives a linear response. Measure the mercury content as described in 3, above. If necessary, carry out several such determinations, using different working standard solutions.

6. Blank Determination

   Carry out all the operations, from ashing to measurement, except for introduction of the sample. When treating the solution, add a quantity of potassium permanganate solution [special reagent (f)] equal to that used for the experimental sample.

7. Calculation

   Read off from the calibration curve the quantities, in μg, of mercury corresponding to the measured absorption values.

   Subtract the quantity of mercury found in the blank from that found in the sample.

   $$\frac{\text{net weight of mercury } (\mu g)}{\text{sample weight (g)}} = \text{ppm Hg}$$

COLOURIMETRIC METHOD FOR DETERMINATION OF MERCURY
(Tentative)

The following suggestion is an amalgamation of the instrumental method for determination of mercury with the colourimetric determination of mercury in iron oxides, which might provide a suitable general "non-instrumental" method of determining mercury.

Procedure

Ash approximately 1 g (accurately weighed) of the sample according to the technique detailed under "Ashing" in the Instrumental Method, using twice the quantities of reagents prescribed. Treat with potassium permanganate, according to "Treatment of the Solution", heat gently to boiling, then allow to cool. Pour the contents of the receiver flask into a separatory funnel (instead of the bubbler used in the instrumental method), wash the receiver flask with water, and add the washings to the contents of the separatory funnel. Then add 1 ml of saturated hydroxylamine hydrochloride which has been brought rapidly to boiling; add, while agitating, 2 ml of a 10% solution of urea. Add slowly a saturated solution of sodium acetate until pH 2 is obtained. Extract the mercury eventually contained in the sample in the separatory funnel by shaking with di-β-naphthylthiocarbazone TS. Use successively 3 ml, 2 ml, and if necessary 1 ml of this TS several times until there is no more colour change produced upon shaking in the separatory funnel. Rinse a last time with pure chloroform.

Combine all extracts in a small test tube and, after eliminating any aqueous phase, empty the contents into a small vial. Then add 4 ml of a 15% solution of sulfuric acid and 2 drops of saturated potassium permanganate solution. Mix well and make certain that the chloroform layer has become completely yellow, which assures the total decomposition of the mercury complex and the quantitative transfer of the mercury into the sulfuric solution. Decant the chloroform layer and eliminate the last traces by washing with 5 ml of petroleum ether.

Pour the sulfuric solution into a curved-neck flask. Wash the petroleum ether phase with 2 ml of 15% sulfuric acid, using also 1 drop of saturated potassium permangante solution and add to the above sulfuric solution. Add 1 ml of saturated potassium permanganate solution and reflux for 30 minutes. Let cool and then add 3.5 ml of sulfuric acid. Heat for 15 minutes more, making sure there is always an excess of permanganate. Allow to cool.

Put the contents of the flask into part B of the steam distillation apparatus (Figure 1). Rinse the cooling system and the flask several times with about 10 ml of water and finally add 3 ml of tin (II) sulfate TS. Connect the various ground-joints of the apparatus as indicated in the scheme, joining "b" and "A" solidly using springs supported on the appropriate spurs. Heat gently flask A, at first half-filled with water, and sheltered from currents of air. Wait to connect the plastic tube "a" between "A" and "b" until a steady jet of steam escapes from "A". Distill through "C", cooling by air, into the micro-Kjeldahl flask in position "D", slightly off-centre, immersed in a dish full of water, and containing 5 ml of an aqueous solution having 10% potassium permanganate and 10% sulfuric acid. Continue the distillation slowly until a volume of distillate is obtained which is 4/5ths that of the micro-Kjeldahl flask. Into the micro-Kjeldahl flask, add 1 ml of saturated hydroxylamine hydrochloride solution, which must, after mixing, bring about a total decolourization. In a second micro-Kjeldahl control flask, place 5 ml of the aqueous solution having 10% potassium permanganate and 10% sulfuric acid, then dilute to a volume equal to that of the distillate contained in the other flask and decolourize by addition of the same amount of hydroxylamine hydrochloride.

./.

Add to the micro-Kjeldahl flask containing the distillate 0.5 ml portions of di-β-naphthylthiocarbazone-chloroform TS until a lilac-mauve colour is obtained. Mix well after each addition and observe the colour in the chloroform layer by decanting it into the cylindrical portion of the flask. Note carefully the volume added.

Add the same amount of di-β-naphthylthiocarbazone-chloroform TS to the control flask. Reduce the blue-green colour of the solvent phase to the same lilac-mauve colour in the first flask by adding, drop by drop, with mixing, a titrated solution of mercury (II) chloride containing 1 μg of mercury per ml. From the volume added, the quantity of mercury present in the test sample can be calculated.

The method described, besides eliminating the causes of error due to individual difference in sensitivity in the perception of the colour and to the interference of traces of mercury possibly present in the reagents, allows the determination with a rigorous specificity, and a precision of at least 5%, of the quantities of mercury of the order of 0.05% μg in absolute value, or 0.05 mg/kg of mercury.

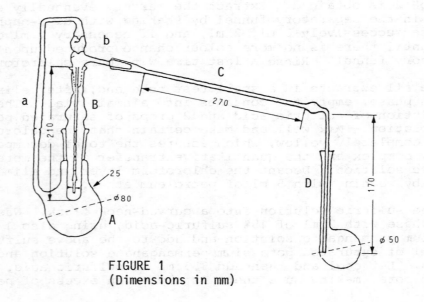

FIGURE 1
(Dimensions in mm)

LIMIT TEST FOR MERCURY

Mercury Detection Instrument

Use any suitable atomic absorption spectrophotometer equipped with a fast-response recorder and capable of measuring the radiation absorbed by mercury vapours at the mercury resonance line of 253.6 nm. A simple mercury vapour meter or detector equipped with a variable span recorder is also satisfactory.

Aeration Apparatus

The apparatus, shown in Fig. 1, consists of a flowmeter (a), capable of measuring at a flow rate of 2.7 l per h, connected via a three-way stopcock (b), with Teflon plug, to 125-ml gas washing bottles (c and d), followed by a drying tube packed with glass wool (e), and finally a suitable quartz liquid absorption cell (f), terminating with a vent (g).

Note: The absorption cell will vary in optical pathlength depending upon the type of mercury detection instrument used.

Bottle c is fitted with an extra-coarse fritted bubbler (Corning 31770 125 EC or equivalent), and the bottle is marked with a 60-ml calibration line. The drying tube e is lightly packed with glass wool or magnesium perchlorate. Bottle c is used for the test solution, and bottle d, which remains empty throughout the procedure, is used to collect water droplets. Alternatively, an apparatus embodying the principle of the assembly described and illustrated may be used. The aerating medium may be either compressed air or compressed nitrogen.

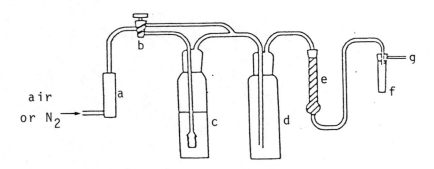

FIGURE 1   Aeration Apparatus

Standard Preparation

Transfer 1.71 g of mercuric nitrate, $Hg(NO_3)H_2O$, to a 1000-ml volumetric flask, dissolve in a mixture of 100 ml of water and 2 ml of nitric acid, dilute to volume with water, and mix. Discard after 1 month. Transfer 10.0 ml of this solution to a second 1000-ml volumetric flask, acidify with 5 ml of dilute sulfuric acid solution (1 in 5), dilute to volume with water, and mix. Discard after 1 week. On the day of use, transfer 10.0 ml of the second solution to a 100-ml volumetric flask, acidify with 5 ml of dilute sulfuric acid (1 in 5), dilute to volume with water, and mix. Each ml of this solution contains 1 µg of Hg. Transfer 2.0 ml of this solution (2 µg of Hg) to a 50-ml beaker, and add 20 ml of water, 1 ml of dilute sulfuric acid solution (1 in 5), and 1 ml of potassium permanganate solution (1 in 25). Cover the beaker with a watch glass, boil for a few seconds, and cool.

Sample Preparation

Prepare as directed in the individual monograph.

Limit Test for Mercury (cont'd)

Procedure

Assemble the aerating apparatus as shown in Figure 1, with bottles c and d empty and stopcock b in the bypass position. Connect the apparatus to the absorption cell (f) in the instrument, and adjust the air or nitrogen flow rate so that, in the following procedure, maximum absorption and reproducibility are obtained without excessive foaming in the test solution. Obtain a baseline reading at 253.6 nm, following the manufacturer's instructions for operating the instrument. Treat the Standard Preparation as follows:

Destroy the excess permanganate by adding a 1-in-10 solution of hydroxylamine hydrochloride, dropwise, until the solution is colourless. Immediately wash the solution into bottle c with water, and dilute to the 60-ml mark with water. Add 2 ml of 10% stannous chloride solution (prepared fresh each week by dissolving 20 g of $SnCl_2 2H_2O$ in 40 ml of warm hydrochloric acid and diluting with 160 ml of water), and immediately reconnect bottle c to the aerating apparatus. Turn stopcock b from the bypass to the aerating position, and obtain the reading on the recorder. Disconnect bottle c from the aerating apparatus, discard the Standard Preparation mixture, wash bottle c with water, and repeat the foregoing procedure using the Sample Preparation; any absorbance produced by the Sample Preparation does not exceed that produced by the Standard Preparation.

DETERMINATION OF WATER CONTENT
(LOSS ON DRYING)

Colours containing $-SO_3Na$ or $-COONa$ groups are usually hygroscopic and any water they retain from their manufacture (or subsequently absorb from the atmosphere) is generally present in the colour in the form of a hydrate. When such colours are dried at 135° the loss in weight may generally be equated to the total water content, but this is not always the case. For example, Erythrosine and Ponceau 4R each retain one molecule of water of crystallization at 135° and it is normal practice to take this into account when totalling the amounts of main components present in a sample.

Procedure

Weigh 2.0 - 3.0 g of the sample in a tared weighing bottle fitted with a ground lid. A weighing bottle of squat form about 50 mm in diameter and 30 mm high is suitable. Heat at the prescribed temperature ± 5° until a constant weight is obtained. Express the loss in weight as a percentage of the weight of sample taken.

DETERMINATION OF CHLORIDE AS SODIUM CHLORIDE

Apparatus

Potentiometric titration apparatus, with silver indicator electrode, calomel reference electrode, and saturated potassium sulfate bridge.

Procedure

Weigh 0.5 – 1.0 g of the dye sample, dissolve in 100 ml of water, and acidify with 5 ml of 1.5 $\underline{N}$ nitric acid solution. Place the silver electrode in the colour solution and connect the calomel electrode to the solution by means of the saturated potassium sulfate bridge. The saturated potassium sulfate bridge may be eliminated by using a glass electrode as the reference electrode; this simplifies the apparatus considerably, and the glass electrode is sufficiently constant to be used as a reference for this type of titration.

Determine the chloride content of the solution by titration against the 0.1 $\underline{N}$ silver nitrate solution, and calculate the result as sodium chloride. 1 ml of 0.1 $\underline{N}$ silver nitrate solution = 0.00585 g of sodium chloride.

Express the result as a percentage of the weight of sample taken.

DETERMINATION OF SULFATE AS SODIUM SULFATE

Weigh 5.0 g of the sample, transfer it to a 250 ml conical flask and dissolve in about 100 ml of water by heating on a water bath. Add 35 g of sulfate-free sodium chloride, stopper the flask, and swirl at frequent intervals during 1 hour. Cool, transfer with saturated sodium chloride solution to a 250 ml measuring flask, and dilute to the mark at 20°. Shake the flask, and filter the solution through a dry filter paper. Pipet 100 ml of the filtrate into a 500 ml beaker, dilute to 300 ml with water and acidify with hydrochloric acid, adding 1 ml in excess. Heat the solution to boiling, and add an excess of 0.25 $\underline{N}$ barium chloride solution, drop by drop, with stirring. Allow the mixture to stand on a hotplate for 4 hours, or leave it overnight at room temperature and then bring it to about 80° and allow the precipitate to settle. Filter off the precipitated barium sulfate, wash with hot water, and ignite at a dull red heat in a tared crucible until a constant weight is obtained.

Carry out a blank determination, apply any necessary correction to the weight of barium sulfate found in the test, and calculate the result as sodium sulfate.

$$\text{Weight of sodium sulfate in sample} = 2.5 \times \text{corrected weight of barium sulfate} \times 0.6086$$

Express the result as a percentage of the weight of sample taken.

DETERMINATION OF SULFATED ASH

Transfer the quantity of the sample directed in the individual monograph to a tared 50- to 100-ml platinum dish or other suitable container, and add sufficient diluted sulfuric acid TS to moisten the entire sample. Heat gently, using a hot plate, an Argand burner, or an infrared heat lamp, until the sample is dry and thoroughly charred, then continue heating until all of the sample has been volatilized or nearly all of the carbon has been oxidized, and cool. Moisten the residue with 0.1 - 0.2 ml of sulfuric acid, and heat in the same manner until the remainder of the sample and any excess sulfuric acid have been volatilized. Finally ignite in a muffle furnace at $800° \pm 25°$ for 15 minutes or longer, if necessary, to complete ignition; cool in a desiccator, and weigh.

NOTE: In order to promote volatilization of sulfuric acid, it is advisable to add a few pieces of ammonium carbonate just before completing ignition.

DETERMINATION OF WATER SOLUBLE CHLORIDES AND SULFATES
IN ALUMINIUM LAKES

Weigh accurately 10 g of the sample. Add 250 ml of water. Stir to wet out the sample and then stir occasionally during a period of 30 minutes. Filter.

Measure 50 ml of the filtrate, add 50 ml water and acidify with 5 ml of 1.5 $\underline{N}$ nitric acid solution. Determine the chloride content by the potentiometric method used for soluble colours.

Measure 50 ml of the filtrate, dilute to 300 ml with water and acidify with hydrochloric acid, adding 1 ml in excess. Heat the solution to boiling and add an excess of 0.25 $\underline{N}$ barium chloride solution, drop by drop, with stirring. Complete the analysis by digesting, filtering, and igniting the precipitate as described in the method used for the determination of sulfate in soluble colours.